Studies in Logic
Volume 50

Intuitionistic Set Theory

Volume 39
Non-contradiction
Lawrence H. Powers, with a Foreword by Hans V. Hansen

Volume 40
The Lambda Calculus. Its Syntax and Semantics
Henk P. Barendregt

Volume 41
Symbolic Logic from Leibniz to Husserl
Abel Lassalle Casanave, ed.

Volume 42
Meta-argumentation. An Approach to Logic and Argumentation Theory
Maurice A. Finocchiaro

Volume 43
Logic, Truth and Inquiry
Mark Weinstein

Volume 44
Meta-logical Investigations in Argumentation Networks
Dov M. Gabbay

Volume 45
Errors of Reasoning. Naturalizing the Logic of Inference
John Woods

Volume 46
Questions, Inferences, and Scenarios
Andrzej Wiśniewski

Volume 47
Logic Across the University: Foundations and Applications. Proceedings of the Tsinghua Logic Conference, Beijing, 2013
Johan van Benthem and Fenrong Liu, eds.

Volume 48
Trends in Belief Revision and Argumentation Dynamics
Eduardo L. Fermé, Dov M. Gabbay, and Guillermo R. Simari

Volume 49
Introduction to Propositional Satisfiability
Victor Marek

Volume 50
Intuitionistic Set Theory
John L. Bell

Studies in Logic Series Editor
Dov Gabbay   dov.gabbay@kcl.ac.uk

Intuitionistic Set Theory

John L. Bell

© Individual author and College Publications 2014, revised 2019.
All rights reserved.

ISBN 978-1-84890-140-7

College Publications
Scientific Director: Dov Gabbay
Managing Director: Jane Spurr

http://www.collegepublications.co.uk

Cover design by Laraine Welch

All rights reserved. No part of this publication may be reproduced, stored in a retrieval system or transmitted in any form, or by any means, electronic, mechanical, photocopying, recording or otherwise without prior permission, in writing, from the publisher.

To Sandra, who has given me new life

Preface

While intuitionistic (or constructive) se theory **IST** has received some attention from mathematical logicians, so far as I am aware no book providing a systematic introduction to the subject has yet been published. This may be the case in part because, as a form of higher-order intuitionistic logic – the internal logic of a topos – **IST** has been chiefly developed in a topos-theoretic context. In particular, proofs of relative consistency with **IST** have been (implicitly) formulated in topos- or sheaf-theoretic terms, rather than in the framework of Heyting-algebra-valued models, the natural extension to **IST** of the well-known Boolean-valued models for classical set theory.

In this book I offer a brief but systematic introduction to **IST** which devops the subject up to and including the use of Heyting-algebra-valued models in relative consistency proofs. I believe that **IST**, presented as it is in the familiar language of set theory, will appeal to those logicians, mathematicians and philosophers who are unacquainted with the methods of topos theory.

The title I originally had in mind for this book was *Constructive Set Theory*. Then it occurred to me that the term "constructive" has come to connote not merely the use of intuitionistic logic, but also the avoidance of impredicative definitions. This is the case, for example, with Aczel's Constructive set theory in which the power set axiom (which permits impredicatve definitions of sets) is not postulated. Since the power set axiom and impredicative definitions are very much a part of **IST**, to avoid confusion I have (with some reluctance) given the book its present title.

JLB January 2014. Typos and other errors corrected December 2018.

Table of Contents

Introduction: Challenging the Logical Presuppositions of Classical Set Theory   1
    The natural numbers and countability   1
    Power sets   5
    The Continuum   7

Chapter I. Intuitionistic Zermelo Set Theory   11
    Axioms and basic definitions   11
    Logical principles in **IZ**   16
    The Axiom of Choice   20

Chapter II. Natural Numbers and Finite Sets   27
    The natural numbers   27
    Models of Peano's axioms   29
    Definitions by recursion   30
    Finite sets   36
    Frege's construction of the natural numbers   40

Chapter III. The Real Numbers   48

Chapter IV. Intuitionistic Zermelo-Fraenkel Set Theory and Frame-Valued Models   55
    Intuitionistic Zermelo-Fraenkel set theory **IZF**   55
    Frame-valued models of **IZF** developed in **IZF**   59
    The consistency of **ZF** and **ZFC** relative to **IZF**   71

Frame-valued models of **IZF** developed in **ZFC**	72
A frame-valued model of **IZF** in which $\mathbb{N}^\mathbb{N}$ is subcountable	79
The Axiom of Choice in frame-valued extensions	83
Real numbers and real functions in spatial extensions	85
Properties of the set of real numbers over $\mathbb{R}$	93
Properties of the set of real numbers over Baire space	95
The independence of the fundamental theorem of algebra from **IZF**	98

Appendix. Heyting Algebras, Frames, and Intuitionistic Logic — 100

Lattices	100
Heyting and Boolean algebras	102
Coverages and their associated frames	107
Connections with logic	108

Concluding Observations — 112

Historical Notes — 115

Bibliography — 117

Index — 121

Introduction

Challenging the Logical Presuppositions of Classical Set Theory

In classical set theory free use is made of the logical principle known as the Law of Excluded Middle (**LEM**): for any proposition p, either p holds or its negation $\neg p$ holds. As we see below, there are a number of intriguing mathematical possibilities which are rendered inconsistent with classical set theory solely as a result of the presence of **LEM**. This suggests the idea of dropping **LEM** in set-theoretical arguments, or, more precisely, basing set theory on *intuitionistic logic*. Accordingly, let us define *Intuitionistic Set Theory* (**IST**) to be any of the usual axiomatic set theories (e.g. Zermelo-Fraenkel set theory **ZF**) based on intuitionistic – rather than classical - logic.

Here are some examples of such mathematical possibilities.

THE NATURAL NUMBERS AND COUNTABILITY.

Call a set *countable* if it is empty or the range of a function defined on the set $\mathbb{N}$ of natural numbers, *subcountable* if it is the range of a function defined on a subset of $\mathbb{N}$, and *numerable* if it is the domain of an injection into $\mathbb{N}$. In classical set theory all of these notions are equivalent, as the following argument shows. Obviously (even in **IST**), every countable set is subcountable. If a set E is subcountable, there is a subset U of $\mathbb{N}$ and a surjection $f\colon U \twoheadrightarrow E$. Then the function $m\colon E \to \mathbb{N}$ defined by $m(x) = $ least $n \in \mathbb{N}$ for which $f(n) = x$ is injective, and it follows that E is numerable. Finally, suppose that E is numerable, and let $m\colon E \rightarrowtail \mathbb{N}$ be an injection. Then (by **LEM**) either $E = \emptyset$ or $E \neq \emptyset$; in the latter case, fix $e \in E$ and define $f\colon \mathbb{N} \to E$ by setting, for $n \in \text{range}(m)$, $f(n) = $ unique $x \in E$ for which $m(x) = n$; and, for $n \notin \text{range}(m)$, $f(n) = e$. Then f is surjective, and so E is countable. It is clear that the validity of this argument rests on two assumptions: **LEM** and the assertion that $\mathbb{N}$ is *well-ordered* (under its natural ordering). Accordingly, if we base our reasoning on intuitionistic logic in which **LEM** is not affirmed, then we can see that, while subcountability obviously continues to be inferable from countability, the argument deriving countability from numerability breaks down because of its dependence on **LEM**. One might suppose that the validity of the argument deriving numerability from subcountability survives the passage to intuitionistic logic, but actually it does not, for it uses the assumption that $\mathbb{N}$ is well-ordered, and this can be shown to *imply* **LEM**. For, given a proposition p, define $U = \{x \in \mathbb{N} : x = 0 \wedge p\} \cup \{1\}$. Notice that $0 \in U \leftrightarrow p$. Then U is nonempty and so, if $\mathbb{N}$ is well-ordered, has a least element n. Since $n \in U$, we have $n = 0 \vee n = 1$. If $n = 0$, then $0 \in U$, whence p; if $n = 1$, then $0 \notin U$, whence $\neg p$. Hence $p \vee \neg p$, and **LEM** follows.

Now, as we have said, it is obvious that any countable set is subcountable, and it is easily shown in **IST** that any numerable set is subcountable. However, in striking contrast with classical set theory, it is consistent with **IST** to assume the existence of sets which are (a) subcountable, but uncountable; (b) numerable, but uncountable; and (c) countable, but not numerable.

Perhaps Cantor's most celebrated theorem is the uncountability of the set $\mathbb{R}$ of real numbers. Cantor first published a proof of this theorem in 1874, but much better known is his second proof, published in 1890, in which he introduces his famous method of "diagonalization". In essence, Cantor's argument establishes that the set $\mathbb{N}^\mathbb{N}$ of all maps $\mathbb{N} \to \mathbb{N}$ is uncountable in the above sense. For given a map $\varphi\colon \mathbb{N} \to \mathbb{N}^\mathbb{N}$, the map $f\colon \mathbb{N} \to \mathbb{N}$ defined by

(*) $$f(n) = \varphi(n)(n) + 1$$

clearly cannot belong to range(φ), so that φ cannot be surjective. This argument does not use **LEM**, and is in fact perfectly valid within **IST**.

Now Cantor would also have accepted the extension of this argument to show that $\mathbb{N}^\mathbb{N}$ cannot be *subcountable* in the above sense. For given $U \subseteq \mathbb{N}$ and a surjection $\varphi\colon U \twoheadrightarrow \mathbb{N}^\mathbb{N}$, if we define $f\colon \mathbb{N} \to \mathbb{N}$ by

(**) $$f(n) = \varphi(n)(n) + 1 \text{ if } n \in U, \ f(n) = 0 \text{ if } n \notin U,$$

then clearly $f \notin \text{range}(\varphi)$ and so again φ fails to be surjective. But this argument uses **LEM** and so is *not* valid within **IST**. In fact, *it is consistent with* **IST** *for* $\mathbb{N}^{\mathbb{N}}$ *to be subcountable,* thus making, oddly, $\mathbb{N}^{\mathbb{N}}$ both subcountable and uncountable. .

The subcountability of $\mathbb{N}^{\mathbb{N}}$ has a number of striking consequences. To begin with, it implies the negation of **LEM**. The simplest way to see this is to note that given $U \subseteq \mathbb{N}$ and a surjection $\varphi: U \twoheadrightarrow \mathbb{N}^{\mathbb{N}}$, the assertion

(*) $$\forall x \in \mathbb{N}[x \in U \vee \neg(x \in U)]$$

is refutable. For if (*) held, then we could extend φ to a surjection $\mathbb{N} \to \mathbb{N}^{\mathbb{N}}$ by assigning the constant value 0 to all $n \notin U$. This would make $\mathbb{N}^{\mathbb{N}}$ countable which we have already shown to be an impossibility. Secondly, U cannot be countable, for the composite of φ with any surjection $\mathbb{N} \twoheadrightarrow U$ would be a surjection $\mathbb{N} \twoheadrightarrow \mathbb{N}^{\mathbb{N}}$, again making $\mathbb{N}^{\mathbb{N}}$, impossibly, countable. Thus U is both uncountable, and as a subset of $\mathbb{N}$, numerable, and it follows that *it is consistent with* **IST** *for* $\mathbb{N}$ *to have an uncountable subset*. Finally, the subcountability of $\mathbb{N}^{\mathbb{N}}$ implies that there is a function defined on a (proper) subset of $\mathbb{N}$ which cannot be extended to the whole of $\mathbb{N}$. To see this, take U and φ as above and define $f: U \to \mathbb{N}$ by setting, for $n \in U$,

$$f(n) = \varphi(n)(n) + 1.$$

Suppose now that f could be extended to a function $g: \mathbb{N} \to \mathbb{N}$. Then since φ is surjective, there is $n_0 \in U$ for which $g = \varphi(n_0)$, leading to the contradiction

$$\varphi(n_0)(n_0) = g(n_0) = f(n_0) = \varphi(n_0)(n_0) + 1.$$

It follows that *it is consistent with* **IST** *there is a function defined on a (proper) subset of* $\mathbb{N}$ *which cannot be extended to the whole of* $\mathbb{N}$.

Now Cantor would also, presumably, have accepted that $\mathbb{N}^{\mathbb{N}}$ cannot be numerable in the sense introduced above. For if $\mathbb{N}^{\mathbb{N}}$ were numerable, then it would (using classical reasoning as before) also have to be countable, contradicting its uncountability. But the argument from numerability to countability does not hold up within **IST** and in fact it is consistent with **IST** for $\mathbb{N}^{\mathbb{N}}$ to be numerable[1].

Let us see what happens (within **IST**) when $\mathbb{N}^{\mathbb{N}}$ is replaced by the set $\mathrm{Par}(\mathbb{N},\mathbb{N})$ of all *partial* functions from $\mathbb{N}$ to $\mathbb{N}$. First, we observe that $\mathrm{Par}(\mathbb{N},\mathbb{N})$ cannot be subcountable, for suppose $U \subseteq \mathbb{N}$ and $\varphi: U \twoheadrightarrow \mathrm{Par}(\mathbb{N},\mathbb{N})$ is a surjection. Let $r \in \mathrm{Par}(\mathbb{N},\mathbb{N})$ be the identity map on $\mathrm{dom}(r) = \{x \in U: x \notin \mathrm{dom}(\varphi(x))\}$. Then since φ is surjective, $r = \varphi(n)$ for some $n \in U$ quickly leading to the contradiction $n \in \mathrm{dom}(r) \leftrightarrow n \notin \mathrm{dom}(r)$.[2]

Nor can $\mathrm{Par}(\mathbb{N}, \mathbb{N})$ be numerable. For suppose $\varphi: \mathrm{Par}(\mathbb{N},\mathbb{N}) \rightarrowtail \mathbb{N}$ is injective, and define $u \in \mathrm{Par}(\mathbb{N},\mathbb{N})$ to be the identity map on

$$\mathrm{dom}(u) = \{x \in \mathbb{N}: \exists f \in \mathrm{Par}(\mathbb{N},\mathbb{N})[\varphi(f) = x \wedge x \notin \mathrm{dom}(f)]\}.$$

Then, writing $n = \varphi(u)$, we have

$$\begin{aligned} n \in \mathrm{dom}(u) &\leftrightarrow \exists f[\varphi(f) = n \wedge n \notin \mathrm{dom}(f)]. \\ &\leftrightarrow \exists f[\varphi(f) = \varphi(u) \wedge n \notin \mathrm{dom}(f)] \\ &\leftrightarrow \exists f[f = u \wedge n \notin \mathrm{dom}(f)] \\ &\leftrightarrow n \notin \mathrm{dom}(u), \end{aligned}$$

a contradiction. So $\mathrm{Par}(\mathbb{N},\mathbb{N})$ is not numerable.

What if we replace $\mathrm{Par}(\mathbb{N},\mathbb{N})$ by the set $\mathrm{Par}^*(\mathbb{N},\mathbb{N})$ of all partial maps on $\mathbb{N}$ with *countable* domains? Suppose that $\mathrm{Par}^*(\mathbb{N},\mathbb{N})$ is actually *countable*, and let $\varphi: \mathbb{N} \twoheadrightarrow \mathrm{Par}^*(\mathbb{N},\mathbb{N})$ be a surjection. If $r \in \mathrm{Par}(\mathbb{N},\mathbb{N})$ is the identity map on $\mathrm{dom}(r) = \{x \in \mathbb{N} : x \notin \mathrm{dom}(\varphi(x))\}$, then the argument above only leads to contradiction when $r \in \mathrm{Par}^*(\mathbb{N},\mathbb{N})$, from which we conclude that

---

[1] See the section **Concluding Observations**.
[2] Clearly this argument continues to hold when $\mathbb{N}$ is replaced by an arbitrary set E.

$r \notin \text{Par}^*(\mathbb{N}, \mathbb{N})$, in other words, $\{x \in \mathbb{N} : x \notin \text{dom}(\varphi(x))\}$ is *uncountable*. In fact, the countability of $\text{Par}^*(\mathbb{N}, \mathbb{N})$ is consistent with **IST**[3].

POWER SETS

Let us turn next to another celebrated theorem of Cantor, namely that, for any set E, the cardinality of E is strictly smaller than that of its power set $\mathbf{P}E$. One way of construing this is the assertion that there can be no surjection $E \to \mathbf{P}E$. Within **IST** this can be proved, as in classical set theory, by employing the argument of Russell's paradox: given $\varphi : E \to \mathbf{P}E$ one defines the "Russell set"

$$R = \{x \in E : x \notin \varphi(x)\}$$

and then shows in the usual way that $R \notin \text{range}(\varphi)$. For $U \subseteq E$, a similar argument, replacing R above by $R \cap U$, shows that there can be no surjection $U \to \mathbf{P}E$. Thus, in particular, within **IST**, $\mathbf{P}\mathbb{N}$ is not subcountable (and so is uncountable).

Equally, the (classically equivalent, but not automatically intuitionistically equivalent) form of Cantor's theorem that, for any set E there is no injection $\mathbf{P}E \rightarrowtail E$ can also be given a proof within **IST** using the idea of Russell's paradox. For suppose given an injection $m: \mathbf{P}E \rightarrowtail E$. Define

$$B = \{x \in E : \exists X \in \mathbf{P}E.\ x = m(X) \wedge x \notin X\}.$$

Writing $m(B) = b$, we have

$$\begin{aligned} b \in B &\leftrightarrow \exists X.\ b = m(X) \wedge b \notin X \\ &\leftrightarrow \exists X.\ m(B) = m(X) \wedge b \notin X \\ &\leftrightarrow \exists X.\ B = X \wedge b \notin X \\ &\leftrightarrow b \notin B, \end{aligned}$$

---

[3] See **Concluding Observations**.

and we obtain our contradiction. In particular, within **IST**, **P**ℕ cannot be numerable.

What if we replace **P**ℕ by the set **P*** ℕ of all *countable* subsets of ℕ? Suppose that **P*** ℕ s actually *countable*, and let φ: ℕ ↠ **P*** ℕ be a surjection. Defining $R = \{x \in \mathbb{N}: x \notin \varphi(x)\}$ as before, the argument above only leads to a contradiction when $R \in$ **P*** ℕ, from which we conclude that $R \notin$ **P*** ℕ, that is, R is *uncountable*. In fact, it follows directly from the consistency of the subcountability of ℕ$^{\mathbb{N}}$ with **IST** that the subcountability of **P***ℕ is also consistent with **IST**[4], since the map ℕ$^{\mathbb{N}}$ → **P*** ℕ : $f \mapsto$ range(f) is surjective.

In classical set theory, **P**E is naturally bijective with 2^E, the set of all maps[5] $E \to 2 = \{0, 1\}$. In **IST**, this is no longer the case. Here, in general, **P**E ≅ Ω^E, where Ω is the object of *truth values or propositions*, that is, the set P1 of all subsets of {∅}. Ω is only identical with 2 when **LEM** is aassumed.[6] In fact, in **IST**, 2^E is isomorphic, not to **P**E, but to its Boolean sublattice **C**E consisting of all *detachable* subsets of E (a subset U of E is said to be detachable if $\forall x \in E(x \in U \vee x \notin U)$). What happens when we replace **P**E by **C**E in the above arguments? Classically, of course, this makes no difference, but do the "Russell's paradox" arguments survive the transition to **IST**? Well, if one takes the first argument, showing that there can be no surjection φ: $E \twoheadrightarrow$ **P**E, one finds that, when **P**E is replaced by **C**E, the set $R \notin$ range(φ) is actually detachable and the argument goes through, proving in **IST** that there can be no surjection $E \to$ **C**E. But the second argument, with **P**E replaced by **C**E (and then E replaced by a subset U of E) goes through in **IST** only if U is detachable. And for the third argument to go through in **IST** once **P**E is replaced by **C**E, it is necessary to show that the set B defined there is detachable. In fact, as we shall see, these arguments can break down completely in **IST** even when E is the set ℝ of real numbers: it is in fact consistent with **IST** that ℝ has just the two detachable subsets ∅, ℝ, so that **C**ℝ (and so also $2^{\mathbb{R}}$) is

---

[4] In fact the countability of P* ℕ is consistent with **IST**, see **Concluding Observations**.
[5] Here we write 0 for ∅ and 1 for {∅}.
[6] On the other hand the natural bijection between Ω^E and **P**E is given, as it is classically, by $f \mapsto f^{-1}(0)$.

isomorphic to 2. *A fortiori* C$\mathbb{R}$ is injectible into $\mathbb{R}$, showing that the third argument fails in **IST**.

THE CONTINUUM

It is characteristic of a continuum that it is "gapless" or "all of one piece", in the sense of not being *actually separated* into two (or more) disjoint nonempty parts. On the other hand, it has been taken for granted from antiquity that continua are *limitlessly divisible,* or *separable* into parts in the sense that any part of a continuum can be "divided", or "separated" into two or more proper parts. Now there is a traditional conceptual difficulty in seeing just how the parts of a continuum obtained by separation—assumed disjoint—"fit together" exactly so as to reconstitute the original continuum. This difficulty is simply illustrated by considering the case in which a straight line X is divided into two segments L, R by cutting it at a point p. What happens to p when the cut is made? On the face of it, there are four possibilities (not all mutually exclusive): (*i*) p is neither in L nor in R; (*ii*) p may be identified as the right-hand endpoint p_L of L: (*iii*) p may be identified as the left-hand endpoint p_R of R; (*iv*) p may be identified as *both* the right-hand endpoint of L *and* the left-hand endpoint of R. Considerations of symmetry suggest that there is nothing to choose between (*ii*) and (*iii*), so that if either of the two holds, then so does the other.

Accordingly we are reduced to possibilities (*i*) and (*iv*). In case (*i*), L and R are disjoint, but since neither contains p, they together fail to cover X; while in case (*iv*), L and R together cover X, but since each contains p, they are not disjoint. This strongly suggests that a (linear) continuum *cannot* be separated, or decomposed, into two disjoint parts *which together cover it.* Herein lies the germ of the idea of cohesiveness.

Of course, this analysis is quite at variance with the account of the (linear) continuum provided by classical set theory. There the continuum is takes the form of the discrete linearly ordered set $\mathbb{R}$ of real numbers. "Cutting" $\mathbb{R}$ (or any interval thereof) at a point p amounts to partitioning it into the pairs of subsets $(\{x: x \leq p\}, \{x: p < x\})$ or $(\{x: x < p\}, \{x: p \leq x\})$: the first and second of these correspond, respectively, to cases (*ii*) and (*iii*) above. Now in the discrete case, one cannot appeal to symmetry as before: consider, for instance, the partitions of the set of natural numbers into the pairs of subsets $(\{n: n \leq 1\}, \{n: 1 < n\})$ and ({n: n < 1}, {n: 1 ≤ n}). The first of these is ({0, 1}, {2, 3, ...}) and the second ({0}, {1, 2, ...}). Here it is manifest that the symmetry naturally arising in the continuous case does not apply: in the first partition 1 is evidently a member of its first component and in the second partition, of its second. In sum, when a discrete linearly ordered set X is "cut", no ambiguity arises as to which segment of the resulting partition the cut point is to be assigned, so that the segments of the partition can be considered disjoint while their union still constitutes the whole of X.

Acknowledging the fact that the set-theoretic continuum, as a discrete entity, *can* be separated into disjoint parts, classical set theory proceeds to capture the characteristic "gaplessness" of a continuum by restricting the *nature* of the parts into which it can be so separated. In set-theoretic topology this is done by confining "parts" to *open* (or *closed*) subsets, leading to the standard topological concept of *connectedness*. Thus a space S is defined to be connected if it cannot be partitioned into two disjoint nonempty open (or closed) subsets — or equivalently, given any partition of S into two open (or closed) subsets, one of the members of the partition must be empty. It is a standard topological theorem that the space $\mathbb{R}$ of real numbers and all of its intervals are connected in this sense.

But now let us return to our original analysis. This led to the idea that a continuum cannot be decomposed into disjoint parts. Let us take the bull by the horns and attempt to turn this idea into a definition. We shall call a space S *cohesive* or *indecomposable*, or a (genuine) *continuum* if, for *any* parts, or subsets U and V of S, whenever $U \cup V = S$ and $U \cap V = \emptyset$, then one of U, V must $= \emptyset$, or, equivalently, one of U, V must $= S$. Clearly S is cohesive precisely when its only detachable subsets are $\emptyset$ and S itself.

Cohesiveness can be furnished with various "logical" formulations. Namely, S is cohesive if and only if, for any property P defined on S, the following implication holds:

(*) $\quad \forall x \in S[P(x) \vee \neg P(x)] \to [\forall x \in S\ P(x) \vee \forall x \in S \neg P(x)]$.

We observe that in classical set theory, the only cohesive spaces are the trivial empty space and one-point spaces. But it turns out that the existence of nontrivial cohesive spaces is consistent with **IST**. In fact it is consistent with **IST** *that $\mathbb{R}$ itself is cohesive.* How does this come about? To get a clue, let us reformulate our definitions in terms of maps, rather than parts. If we denote by **2** the two-element discrete space, then connectedness of a space S is equivalent to the condition that any *continuous* map $S \to \mathbf{2}$ is constant, and cohesiveness of S to the condition that any map $S \to \mathbf{2}$ *whatsoever* is constant. Supposing S to be connected and to possess more than one point, then from **LEM** it follows that there exist nonconstant—and hence discontinuous— maps $S \to \mathbf{2}$. But the situation would be decidedly otherwise if *all* maps defined on S were continuous, for then, clearly, the connectedness of S would immediately yield its cohesiveness. In fact it is consistent with **IST** that all maps $\mathbb{R} \to \mathbb{R}$ - and hence all maps $\mathbb{R} \to \mathbf{2}$ - are continuous. It follows that it is consistent with **IST** that $\mathbb{R}$ is cohesive.

The consistency with **IST** of all these possibilities can be established by constructing models of **IST** in which they can be shown to hold. In this book we shall establish both the consistency with **IST** of the subcountability of $\mathbb{N}^\mathbb{N}$ and the cohesiveness of $\mathbb{R}$ through the use of *Heyting-algebra valued* models, of which the more familiar Boolean-valued models of classical set theory are special cases.

*

Some years ago Paul Cohen published an article in the *Scientific American* entitled *Non- Cantorian Set Theory.* There he described the set theories in which Cantor's continuum hypothesis is violated. In terming these set theories non-Cantorian he was making an analogy with non-Euclidean geometries in which the parallel postulate is violated. If, in Cohen's analogy, non–Cantorian set theories correspond to non-Euclidean geometries, then classical (Zermelo-Fraenkel) set theory corresponds to *neutral* or *absolute* geometry in which no form of the parallel postulate is laid down. Let us reformulate Cohen's analogy by replacing the continuum hypothesis with the Law of Excluded Middle; *it is then intuitionistic set theory that corresponds to neutral geometry.* Intuitionistic set theory can thus be seen as a "neutral" set theory, compatible with a number of principles – such as the subcountability of $\mathbb{N}^\mathbb{N}$ and the cohesiveness of the real line - which are incompatible with classical set theory.

Janos Bolyai, one of the inventors/discoverers of non-Euclidean geometry, was moved to describe it as a "strange new universe" - and indeed it was, by the canons of Euclidean geometry. Similarly, certain of the various extensions of intuitionistic set theory described above may strike one as "strange new universes" in comparison with the familiar universe of classical set theory. But, just as geometers became familiar with non-Euclidean geometry, providing it with models (such as the pseudosphere) which made it seem "natural", so seemingly curious properties compatible with intuitionistic set theory - such as the subcountability of $\mathbb{N}^\mathbb{N}$ or the cohesiveness of the real line - become clear when their meanings in the models realizing them are grasped.

In this book we shall formulate and develop versions of intuitionistic Zermelo- and Zermelo-Fraenkel set theories – **IZ** and **IZF**, respectively, and construct the Heyting-algebra valued models which will be used to establish the relative consistency with **IZF** of some of the assertions we have discussed above.

Chapter I

Intuitionistic Zermelo Set Theory

AXIOMS AND BASIC DEFINITIONS

Intuitionistic set theory is formulated as a system of axioms in the same first-order language as its classical counterpart, only based on intuitionistic logic. The *language of set theory* is a first-order language $\mathcal{L}$ with equality, which includes a binary symbol $\in$. We write $x \neq y$ for $\neg (x = y)$ and $x \notin y$ for $\neg (x \in y)$. Individual variables $x, y, z, ...$ of $\mathcal{L}$ will be understood as ranging over *sets*. The unique existential quantifier $\exists!$ is introduced by writing, for any formula $\varphi(x)$, $\exists!x\varphi(x)$ as an abbreviation of the formula $\exists x[\varphi(x) \wedge \forall y(\varphi(y) \to x = y)]$.

$\mathcal{L}$ will also allow the formation of terms of the form $\{x: \varphi(x)\}$, for any formula φ containing the free variable x.. Such terms are called *classes*; we shall use upper case letters $A, B, ...$ for classes. For each class $A = \{x: \varphi(x)\}$ the formula

$$\forall x[x \in A \leftrightarrow \varphi(x)]$$

is called the *defining axiom* for the class A. Two classes A, B are defined to be *equal*. and we write $A = B$ if

$$\forall x(x \in A \leftrightarrow x \in B).$$

A is a *subclass* of B, and we write $A \subseteq B$, if

$$\forall x(x \in A \to x \in B).$$

We also write **Set**(A) for the formula

$$\exists u \forall x(x \in A \leftrightarrow x \in u).$$

Set(A) asserts that the class A is a set. For any set u, it follows from the defining axiom for the class $\{x: x \in u\}$ that **Set**($\{x: x \in u\}$). We shall identify $\{x: x \in u\}$ with u, so that sets may be considered as (special sorts of) classes and we may introduce assertions such as $u \subseteq A$, $u = A$, etc.

If A is a class, we write $\exists x \in A \varphi(x)$ for $\exists x(x \in A \wedge \varphi(x))$ and $\forall x \in A \varphi(x)$ for $\forall x(x \in A \to \varphi(x))$.

We define the following classes:

- $\{u_1,\ldots,u_n\} = \{x : x = u_1 \vee \ldots \vee x = u_n\}$
- $\bigcup A = \{x : \exists y (y \in A \wedge x \in y)\}$
- $\bigcap A = \{x : \forall y (y \in A \to x \in y)\}$
- $A \cup B = \{x : x \in A \vee x \in B\}$
- $A \cap B = \{x : x \in A \vee x \in B\}$
- $A - B = \{x : x \in A \wedge x \notin B\}$
- $u^+ = u \cup \{u\}$
- $PA = \{x: x \subseteq A\}$
- $\{x \in A: \varphi(x)\} = \{x: x \in A \wedge \varphi(x)\}$
- $V = \{x: x = x\}$
- $\emptyset = \{x: x \neq x\}$

The system **IZ** of *intuitionistic Zermelo set theory* is based on the following axioms:

Extensionality $\forall u \, \forall v [\forall x(x \in u \leftrightarrow x \in v) \to u = v]$

Empty Set   **Set**($\emptyset$)

Pairing $\forall u \, \forall v \,$**Set**($\{u, v\}$)

Union $\forall u \,$**Set**($\bigcup u$)

Powerset $\forall u \,$**Set**(Pu)

Infinity $\exists a[\emptyset \in a \wedge \forall x \in a(x^+ \in a)]$

Separation $\forall u_1 .... \forall u_n \forall a \; \mathbf{Set}(\{x \in a : \varphi(x, u_1, ..., u_n)\})$

Until further notice all propositions, theorems, etc. will be proved in **IZ** (using the above axioms and intuitiuonistic logic[7]).

Let $\varphi(x)$ be a formula of $\mathcal{L}$ and $t(x)$ be a term of $\mathcal{L}$ such that the sentence $\forall x \; \mathbf{Set}(t(x))$ is provable in **IZ**. Then we write $\{t(x): \varphi(x)\}$ for the class

$$\{y: \exists x. \; y = t(x) \wedge \varphi(x)\}.$$

We also write $\bigcup_{\varphi(x)} t(x)$ for the class

$$\{y: \exists x. \; y \in t(x) \wedge \varphi(x)\}$$

and $\bigcap_{\varphi(x)} t(x)$ for the class

$$\{y: \forall x (\varphi(x) \to y \in t(x))\}.$$

Because we are using intuitionistic logic, we must distinguish carefully between the assertions $A \neq \emptyset$ (A is *nonempty*) and $\exists x. \; x \in A$ (A is *inhabited*). While an inhabited set is nonempty, the converse does not hold in general.

We write 0 for $\emptyset$, 1 for $\{0\}$ and 2 for $\{0, 1\}$. 2 carries the natural ordering $\leq$ given by $0 \leq 0, 0 \leq 1, 1 \leq 1$.

The *ordered pair* of two sets u, v is defined as usual by

$$\langle u, v \rangle = \{\{u\}, \{u, v\}\}.$$

Clearly we have

---

[7] For an account of intuitionistic logic, seed the Appendix.

$$\forall u \forall v \, \mathbf{Set}(\langle u,v \rangle).$$

Proposition 1. $\quad <u, v> = <a, b> \rightarrow u = a \wedge v = b.$

Proof. This must be proved without using **LEM**. Suppose that $<u, v> = <a, b>$.

Since $\{u\}$ is an element of $<u, v>$, it must also be an element of $<a, b>$, so that either $\{u\} = \{a\}$ or $\{u\} = \{a, b\}$. In both cases $u = a$.

Since $\{u, v\}$ is an element of $<u, v>$, it must also be an element of $<a, b>$, so that either $\{u, v\} = \{a\}$ or $\{u, v\} = \{a, b\}$. In either case $v = a$ or $v = b$. If $v = a$ then $u = a = v$, so that

$$\{\{a\}\} = \{\{u\}, \{u, v\}\} = <u, v> = <a, b> = \{\{a\}, \{a, b\}\}.$$

It follows that $\{a\} = \{a, b\}$ so that $a = b$, and so $v = b$. So in either case $v = b$, and the proposition is proved. ∎

We define the *Cartesian product* of two classes A and B by

$$A \times B = \{< x, y >: x \in A \wedge y \in B\}.$$

It is left to the reader as an exercise to show that, **Set**(A) and **Set**(B) implies **Set**(A × B)

A (binary) *relation* between classes A, B is a subset $R \subseteq A \times B$. We sometimes write aRb for $<a, b> \in R$. The *doman* dom(R) and the *range* ran(R) of R are defined by

$$\text{dom}(R) = \{x: \exists y \, xRy\} \quad \text{ran}(R) = \{y: \exists x \, xRy\}.$$

It is left as an exercise to the reader to show that, if **Set**(R), then **Set**(dom(R)) and **Set**(ran(R)).

A relation F is a *function*, or *map*, written Fun(F), if for each $a \in \mathrm{dom}(F)$ there is a unique b for which aFb. This unique b is written $F(a)$ or Fa. We write[8] $F: A \to B$ for the assertion that F s a function with $\mathrm{dom}(F) = A$ and $\mathrm{ran}(F) = B$. In this case we occasionally write $a \mapsto F(a)$ for F.

The *identity map* 1_A on A is the map $A \to A$ given by $a \mapsto a$. If $X \subseteq A$, the map $x \mapsto x: X \to A$ is called the *insertion map* of X into A.

If $F: A \to B$ and $X \subseteq A$, the *restriction* $F|X$ of F to X s the map $X \to A$ given by $x \mapsto F(x)$. If $Y \subseteq B$, the *inverse image* of Y under F is the set

$$F^{-1}[Y] = \{x \in A: F(x) \in Y\}.$$

Given two functions $F: A \to B$, $G: B \to C$, we define the *composite function* $G \circ F: A \to C$ to be the function $a \mapsto G(F(a))$. If $F: A \to A$, we write F^2 for $F \circ F$, F^3 for $F \circ F \circ F$ etc.

A function $F: A \to B$ is said to be *monic* if for all $x, y \in A$, $F(x) = F(y)$ implies $x = y$, *epi* if for any $b \in B$ there is $a \in A$ for which $b = F(a)$, and *bijective*, or a *bijection*, if it is both monic and epi. It is easily shown that F is bijective if and only if F has an *inverse*, that is, a map $G: B \to A$ such that $F \circ G = 1_B$ and $G \circ F = 1_A$. Two sets X and Y are said to be *equipollent*, and we write $X \approx Y$, if there is a bijection between them.

Suppose we are given two classes I, A and an epi map $F: I \to A$. Then $A = \{F(i): i \in I\}$ and so, if, for each $i \in I$, we write a_i for $F(i)$, then A can be presented in the form of an *indexed class* $\{a_i: i \in I\}$. If A is presented as an indexed class of sets $\{X_i: i \in I\}$, then we write $\bigcup_{i \in I} X_i$ and $\bigcap_{i \in I} X_i$ for $\bigcup A$ and $\bigcap A$, respectively.

The *projection maps* $\pi_1: A \times B \to A$ and $\pi_2: A \times B \to B$ are defined to be the maps $\langle a, b \rangle \mapsto a$ and $\langle a, b \rangle \mapsto b$ respectively.

---

[8] The bold arrow ($\to$) here is not to be confused with the arrow ($\to$) for implication.

For sets A, B, the *exponential* B^A is defined to be the set of all functions from A to B. (Exercise: show that this is indeed a set.)

It follows easily from the axioms and definitions of **IZ** that, for any set A, PA, under the partial ordering $\subseteq$, is a frame[9] with operations $\cup, \cap$ and $\Rightarrow$, where

$$U \Rightarrow V = \{x : x \in U \to x \in V\},$$

Its top and bottom elements are A and $\emptyset$ respectively.

For any set a, we write $\{a \mid \varphi\}$ for $\{x : x = a \wedge \varphi\}$; notice that

$$a \in \{a \mid \varphi\} \leftrightarrow \varphi.$$

From Extensionality we infer that $\{a \mid \varphi\} = \{a \mid \psi\}$ iff $(\varphi \leftrightarrow \psi)$; thus, in particular, the elements of $P1$ (recall that $1 = \{0\}$) correspond naturally to *truth values*, i.e. propositions identified under equivalence. $P1$ is called the *frame of truth values* and is denoted by Ω. The top element 1 of Ω is usually written *true* and the bottom element $\emptyset$ as *false*.

In **IZ**, Ω plays the role of a *subset classifier*. That is, for each set A, subsets of A are correlated bijectively with functions $A \to \Omega$. To wit, each subset $X \subseteq A$ is correlated with its *characteristic function* $\chi_X : A \to \Omega$ given by $\chi_X(x) = \{0 \mid x \in X\}$; conversely each function $f : A \to \Omega$ is correlated with the subset $f^{-1}(1) = \{x \in A : f(x) = 1\}$ of A.

LOGICAL PRINCIPLES IN IZ

Properties of Ω correspond to *logical principles* of the set theory. For instance, consider the logical principles (where φ, ψ are any formulas):

    LEM (law of excluded middle) $\varphi \vee \neg \varphi$
    WLEM (weakened law of excluded middle) $\neg \varphi \vee \neg \neg \varphi$.
    DML (De Morgan's law) $\neg(\varphi \wedge \psi) \to (\neg \varphi \vee \neg \psi)$

---

[9] Frames are defined in the Appendix.

In intuitionistic logic **WLEM** and **DML** are equivalent.

LEM and **WLEM** correspond respectively to the properties

$$\forall \omega \in \Omega.\ \omega = \textit{true} \vee \omega = \textit{false} \qquad \forall \omega \in \Omega.\ \omega = \textit{false} \vee \omega \neq \textit{false}.$$

Given a formula $\varphi(x, y)$, the sentence $\forall x \forall y (\varphi \vee \neg \varphi)$ will be read as asserting that φ is *decidable*. For a class A, the sentence $\forall x \in A\ \forall y \in A\ (x = y \vee x \neq y)$ will be read as asserting that A is *discrete*. We then have

Proposition 2. In IZ, each of the following is equivalent to **LEM**:

(i) *Membership is decidable*, i.e. $\forall x \forall y\ (x \in y \vee x \notin y)$
(ii) V *is discrete*
(iii) *Every set is discrete*
(iv) Ω *is discrete*
(v) $\Omega = 2$
(vi) $\forall x\ (0 \in x \vee 0 \notin x)$
(vii) $(2, \leq)$ *is well-ordered, i.e. every inhabited subset of 2 has a $\leq$–least element.*

Proof. That **LEM** implies (i) is obvious.

(i) $\rightarrow$ (ii). Assuming (i), we have $\forall x \forall y\ (x \in \{y\} \vee x \notin \{y\})$, whence (ii).

(ii) $\rightarrow$ (iii) and (iii) $\rightarrow$ (iv) are both obvious.

(iv) $\rightarrow$ (v). Assuming (iv), for any $\omega \in \Omega$ we have $\omega = \{0\}$ or $\omega \neq \{0\}$. In the latter case $\omega = \emptyset$, and (v) follows.

(v) $\rightarrow$ (vi). Let φ be the formula $0 \in x$. Then, assuming (v), we have $\{0 \mid \varphi\} = \{0\}$ or $\{0 \mid \varphi\} = \emptyset$. In the first case we get φ and in the second $\neg \varphi$. Hence (vi).

(vi) $\rightarrow$ (vii). Let U be an inhabited subset of 2. Assuming (vi), we have $0 \in U$ or $0 \notin U$. In the first case 0 is the least element of U. In the second case, since U is inhabited we must have $1 \in U$, so that 1 is the least element of U. Thus in either case U has a least element, and (vii) follows.

(viii) → **LEM**. Assume (vii) and let φ be any formula. Let U be the subset $\{0 \mid \varphi\} \cup \{1\}$ of 2. Then $0 \in U \leftrightarrow \varphi$; moreover U is inhabited and so has a least element a, which must be either 0 or 1. If $a = 0$, then $0 \in U$, whence φ; while if $a = 1$, then $0 \notin U$, whence $\neg \varphi$. **LEM** follows. ∎

Observe that the negation operation $\neg$ on formulas corresponds to the complementation operation on Ω; we use the same symbol $\neg$ to denote the latter. This operation satisfies (using ω, ω' as variables ranging over Ω)

$$\omega \subseteq \neg \omega' \leftrightarrow \omega \cap \omega' = \text{false}.$$

Classically, $\neg$ also satisfies the dual law, viz.

$$\neg \omega \subseteq \omega' \leftrightarrow \omega \cup \omega' = \text{true}.$$

But in **IZ**, this is far from being the case. In fact we can prove

Proposition 3. *In **IZ**, **LEM** is equivalent to the assertion that there exists an operation* $-: \Omega \to \Omega$ *satisfying*

(*) $\qquad\qquad\qquad -\omega \subseteq \omega' \leftrightarrow \omega \cup \omega' = \text{true}$

Proof. If **LEM** holds, then the complementation operation $\neg$ satisfies (*).

Conversely, suppose given an operation $-: \Omega \to \Omega$ for which (*) holds. Then

$$-\text{true} \subseteq \text{false} \leftrightarrow \text{false} \cup \text{true} = \text{true},$$

so that $-\text{true} \subseteq \text{false}$, whence $-\text{true} = \text{false}$. Next,

$$0 \in -\omega \wedge 0 \in \omega \to 0 \in -\omega \wedge \omega = \text{true} \to 0 \in -\text{true} = \text{false}.$$

Since $0 \notin \text{false}$, it follows that

$$0 \in -\omega \to 0 \notin \omega \to 0 \in \neg \omega,$$

and from this we infer that $-\omega \subseteq \neg\omega$. Since, obviously, $\omega \cup -\omega = \text{\textit{true}}$, it then follows that, for any ω, $\omega \cup \neg\omega = \text{\textit{true}}$, which is **LEM**. ∎

The weak law of excluded middle is also equivalent to a certain property of the ordered set $(2, \leq)$:

Proposition 4. *In* **IZ**, *the following are equivalent*:

(i) WLEM
(ii) $(2, \leq)$ *is complete, i.e. every subset of 2 has a $\leq$-supremum.*

Proof. (i) → (ii). Assume **(i)** and let $U \subseteq 2$. Clearly $1 \notin U \leftrightarrow U \subseteq \{0\}$. By **(i)**, we have $1 \notin U$ or $\neg(1 \notin U)$. In the first case $U \subseteq \{0\}$ and U then has supremum 0. In the second case $\neg(U \subseteq \{0\})$, so that $\neg\forall x \in U . x = 0$, which is equivalent to

$$\forall x \in U (x \leq 0) \to 0 = 1.$$

Also, obviously,

$$\forall x \in U (x \leq 1) \to 1 = 1.$$

It follows that, for any $u \in 2$,

$$\forall x \in U . x \leq u \to u = 1,$$

so that U has supremum 1. Thus in either case U has a supremum, and **(ii)** follows.

(ii) → (i). Assume **(ii)**, let φ be any formula and define $U = \{1 : \varphi\}$. Then U has a supremum a and there are two cases: $a = 0$ or $a = 1$. In the first case $1 \notin U$, so that $\neg\varphi$. In the second case, if $\neg\varphi$, then $U = \emptyset$ and so $a = 0$, which is impossible. Therefore $\neg\neg\varphi$, and **(i)** follows. ∎

THE AXIOM OF CHOICE

A *choice function* on a set A is a function f with domain A such that $f(a) \in a$ whenever a is inhabited. The *Axiom of Choice* **AC** is the assertion that every set has a choice function. While **AC** plays a major role in classical set theory, in an intuitionistic setting it is far too strong, since even very weak versions of it can be shown to imply **LEM**. In fact we have

Proposition 5. *It is provable in* **IZ** *that if each doubleton has a choice function, then* **LEM** *holds (and, of course, conversely).*

Proof. Let φ be any formula; define $U = \{x \in 2 : x = 0 \vee \varphi\}$ and $V = \{x \in 2 : x = 1 \vee \varphi\}$. Suppose given a choice function f on $\{U, V\}$. Writing $a = f(U)$, $b = f(V)$, we then have $a \in U$, $b \in V$, i.e.

$$(a = 0 \vee \varphi) \wedge (b = 1 \vee \varphi).$$

Hence

$$(a = 0 \wedge b = 1) \vee \varphi,$$

whence

(*) $$a \neq b \vee \varphi.$$

But

$$\varphi \to U = V \to a = b,$$

so that

$$a \neq b \to \neg \varphi.$$

This, together with (*), gives $\varphi \vee \neg \varphi$. ∎

As we have seen, in **IZ** the Law of Excluded Middle is derivable from **AC**. We are now going to show that each of a number of classically correct, but intuitionistically invalid logical principles, including the Law of Excluded Middle for sentences, is, in **IZ**, equivalent to a suitably *weakened* version of **AC**. Thus each of these logical principles may be viewed as a choice principle.

We fix some more notation. For each set A we shall write QA for the set of *inhabited* subsets of A, that is, of subsets X of A for which $\exists x\,(x \in X)$. The class of functions with domain A will be denoted by $\text{Fun}(A)$.

We tabulate the following new *logical schemes*

- **SLEM** $\alpha \vee \neg\alpha$
- **Lin** $(\alpha \to \beta) \vee (\beta \to \alpha)$
- **SWLEM** $\neg\alpha \vee \neg\neg\alpha$
- **Ex**[10] $\exists x[\exists x\varphi(x) \to \varphi(x)]$
- **Un** $\exists x[\varphi(x) \to \forall x\varphi(x)]$
- **Dis**[11] $\forall x[\alpha \vee \varphi(x)] \to \alpha \vee \forall x\varphi(x)$

Here α and β are sentences, and $\varphi(x)$ is a formula with free variable x. In intuitionistic logic, **Lin** and **SWLEM** are consequences of **SLEM**; and **Un** implies **Dis**. All of these schemes follow, of course, from **LEM**, the full Law of Excluded Middle.

We formulate the following *choice principles* — here X is an arbitrary set and $\theta(x, y)$ an arbitrary formula with at most the free variables x, y:

- $\mathbf{AC}_X$ $\forall x \in X\ \exists y\ \theta(x, y) \to \exists f \in \text{Fun}(X)\ \forall x \in X\ \theta(x, fx)$
- $\mathbf{AC}^*_X$ $\exists f \in \text{Fun}(X)\ [\forall x \in X\ \exists y\ \theta(x, y) \to \forall x \in X\ \theta(x, fx)]$
- $\mathbf{DAC}_X$ $\forall f \in \text{Fun}(X)\ \exists x \in X\ \theta(x, fx) \to \exists x \in X\ \forall y\ \theta(x, y)$
- $\mathbf{DAC}^*_X$ $\exists f \in \text{Fun}(X)\ [\exists x \in X\ \theta(x, fx) \to \exists x \in X\ \forall y\ \theta(x, y)]$

---

[10] In intuitionistic logic **Ex** is equivalent to the *independence of premises rule*:
$$\frac{\alpha \to \exists x\varphi(x)}{\exists x\,(\alpha \to \varphi(x))}$$

[11] In intuitionistic logic **Dis** is equivalent to the *higher dual distributive law*
$$\forall x[\alpha(x) \vee \beta(x)] \to \exists x\alpha(x) \vee \forall x\beta(x).$$

The first two of these are forms of the Axiom of Choice for X; while classically equivalent, in **IZ** **AC*** $_X$ implies **AC**$_X$, but not conversely. The principles **DAC**$_X$ and **DAC**$^*_X$ are *dual* forms of the Axiom of Choice for X: classically they are both equivalent to **AC**$_X$ and **AC**$^*_X$, but in **IZ** **DAC**$^*_X$ implies **DAC**$_X$, and not conversely.

We also formulate the *weak extensional selection principle*:

WESP $\quad \exists x \in 2\ \varphi(x) \wedge \exists x \in 2\ \psi(x) \to$
$$\exists x \in 2 \exists y \in 2 [\varphi(x) \wedge \psi(y) \wedge [\forall x \in 2[\varphi(x) \leftrightarrow \psi(x)] \to x = y]].$$

Here $\varphi(x)$, $\psi(x)$ are formulas with the free variable x. This principle asserts that, for any pair of instantiated properties of members of 2, instances may be assigned to the properties in a manner that depends just on their extensions. **WESP** is a straightforward consequence of **AC**$_{Q2}$. For taking $\theta(u, y)$ to be $y \in u$ in **AC**$_{Q2}$ yields the existence of a function f with domain Q2 such that $fu \in u$ for every $u \in Q2$. Given formulas $\varphi(x)$, $\psi(x)$, and assuming the antecedent of **WESP**, the sets $U = \{x \in 2: \varphi(x)\}$ and $V = \{x \in 2: \psi(x)\}$ are members of Q2, so that $a = fU \in U$, and $b = fV \in V$, whence $\varphi(a)$ and $\psi(b)$. Also, if $\forall x \in 2[\varphi(x) \leftrightarrow \psi(x)]$, then $U = V$, whence $a = b$; it follows then that the consequent of **WESP** holds.

We show that each of the logical principles tabulated above is equivalent (over **IZ**) to a choice principle. Starting at the top of the list, we have first:

Proposition 6. **WESP** *and* **SLEM** *are equivalent over* **IZ**.

Proof. Assume **WESP**. Let α be any sentence and define

$$\varphi(x) \equiv x = 0 \vee \alpha \qquad \psi(x) \equiv x = 1 \vee \alpha.$$

With these instances of φ and ψ the antecedent of **WESP** is clearly satisfied, so that there exist members a, b of 2 for which (1) $\varphi(a) \wedge \psi(b)$ and (2) $\forall x [[\forall x \in 2[\varphi(x) \leftrightarrow \psi(x)] \to a = b$. It follows from (1) that $\alpha \vee (a = 0 \wedge b = 1)$, whence (3) $\alpha \vee a \neq b$. And since clearly $\alpha \to \forall x \in 2[\varphi(x) \leftrightarrow \psi(x)]$ we deduce from (2) that $\alpha \to a = b$, whence $a \neq b \to \neg \alpha$. Putting this last together with (3) yields $\alpha \vee \neg \alpha$, and **SLEM** follows.

For the converse, we argue informally. Suppose that **SLEM** holds. Assuming the antecedent of **WESP**, choose $a \in 2$ for which $\varphi(a)$. Now (using **SLEM**) define an element $b \in 2$ as follows. If $\forall x \in 2[\varphi(x) \leftrightarrow \psi(x)]$ holds, let $b = a$; if not, choose b so that $\beta(b)$. It is now easy to see that a and b satisfy $\varphi(a) \wedge \psi(b) \wedge [\forall x \in 2[\varphi(x) \leftrightarrow \psi(x)] \to a = b]$. **WESP** follows. ∎

Next, we observe that, while $\mathbf{AC}_1$ is (trivially) provable in **IZ**, by contrast we have

Proposition 7. $\mathbf{AC}_1^*$ and **Ex** are equivalent over **IZ**.

Proof. Assuming $\mathbf{AC}_1^*$, take $\theta(x, y) \equiv \varphi(y)$ in its antecedent. This yields an $f \in \mathrm{Fun}(1)$ for which $\forall y \varphi(y) \to \varphi(f0)$, giving $\exists y[\exists y \varphi(y) \to \varphi(y)]$, i.e., **Ex**.

Conversely, define $\varphi(y) \equiv \theta(0, y)$. Then, assuming **Ex**, there is b for which $\exists y \varphi(y) \to \varphi(b)$, whence $\forall x \in 1\, \exists y \theta(x, y) \to \forall x \in 1\, \theta(x, b)$. Defining $f \in \mathrm{Fun}(1)$ by $f = \{\langle 0, b \rangle\}$ gives $\forall x \in 1\, \exists y \theta(x, y) \to \forall x \in 1\, \theta(x, fx)$, and $\mathbf{AC}_1^*$ follows. ∎

Further, while $\mathbf{DAC}_1$ is easily seen to be provable in **IZ**, we have

Proposition 8. $\mathbf{DAC}_1^*$ and **Un** are equivalent over **IZ**.

Proof. Given φ, Define $\theta(x, y) \equiv \varphi(y)$. Then, for $f \in \mathrm{Fun}(1)$, $\exists x \in 1\, \theta(x, fx) \leftrightarrow \varphi(f0)$ and $\exists x \in 1\, \forall y \theta(x, y) \leftrightarrow \forall y \varphi(y)$. $\mathbf{DAC}_1^*$ then gives

$$\exists f \in \mathrm{Fun}(1)[\varphi(f0) \to \forall y \varphi(y)],$$

from which **Un** follows easily.

Conversely, given φ, define $\varphi(y) \equiv \theta(0, y)$. Then from **Un** we infer that there exists b for which $\varphi(b) \to \forall y \varphi(y)$, i.e. $\theta(0, b) \to \forall y \theta(0, y)$. Defining $f \in \mathrm{Fun}(1)$ by $f = \{\langle 0, b \rangle\}$ then gives $\theta(0, f0) \to \exists x \in 1\, \forall y \theta(x, y)$, whence $\exists x \in 1\, \theta(x, fx) \to \exists x \in 1\, \forall y \theta(x, y)$, and **Un** follows. ∎

Next, while **AC₂** is easily proved in **IZ**, by contrast we have

Proposition 9. **DAC₂** *and* **Dis** *are equivalent over* **IZ**.

Proof. The antecedent of **DAC₂** is equivalent to the assertion

$$\forall f \in \text{Fun}(2)[\theta(0, f0) \lor \theta(1, f1)],$$

which, in view of the natural correlation between members of Fun (2) and ordered pairs, is equivalent to the assertion

$$\forall y \forall y'[\theta(0, y) \lor \theta(1, y')].$$

The consequent of **DAC₂** is equivalent to the assertion

$$\forall y \in Y \theta(0, y) \lor \forall y' \in Y \theta(1, y')$$

So **DAC₂** itself is equivalent to

$$\forall y \forall y'[\theta(0,y) \lor \theta(1,y')] \;\to\; \forall y \theta(0,y) \lor \forall y' \theta(1,y').$$

But this is obviously equivalent to the scheme

$$\forall y \forall y'[\varphi(y) \lor \psi(y')] \;\to\; \forall y \varphi(y) \lor \forall y' \psi(y'),$$

where y does not occur free in ψ, nor y' in φ. And this last is easily seen to be equivalent to **Dis**. ∎

Now consider **DAC₂***. This is quickly seen to be equivalent to the assertion

$$\exists z \exists z'[\theta(0, z) \lor \theta(1, z') \;\to\; \forall y \theta(0, y) \lor \forall y' \theta(1, y'),$$

i.e. to the assertion, for arbitrary $\varphi(x), \psi(x)$, that

$$\exists z \exists z'[\varphi(z) \vee \psi(z')] \to \forall y \varphi(y) \vee \forall y' \beta(y')].$$

This is in turn equivalent to the assertion, for any sentence α,

(*) $\exists y[\alpha \vee \psi(y)] \to \alpha \vee \forall y \psi(y)]$.

Now (*) obviously entails **Un**. Conversely, given **Un**, there is b for which $\psi(b) \to \forall y \psi(y)$. Hence $\alpha \vee \psi(b) \to \alpha \vee \forall y \psi(y)$, whence (*). So we have proved

Proposition 10. *Over* **IZ**, $\mathbf{DAC}_2^*$ *is equivalent to* **Un**, *and hence also to* $\mathbf{DAC}_1^*$. ∎

In order to provide choice schemes equivalent to **Lin** and **Stone** we introduce

$\mathbf{ac}_X^*$ $\quad \exists f \in 2^X [x \in X \exists y \in 2 \, \theta(x, y) \to \forall x \in X \, \theta(x, fx)]$

$\mathbf{wac}_X^*$ $\quad \exists f \in 2^X [\forall x \in X \exists y \in 2 \, \theta(x, y) \to \forall x \in X \, \theta(x, fx)]$ provided the sentence $\forall x[\theta(x, 0) \to \neg \theta(x, 1)]$ is provable in **IZ**.

Clearly $\mathbf{ac}_X^*$ is equivalent to

$$\exists f \in 2^X [\forall x \in X[\theta(x, 0) \vee \theta(x, 1)] \to \forall x \in X \, \theta(x, fx)]$$

and similarly for $\mathbf{wac}_X^*$. Then we have

Proposition 11. *Over* **IZ**, $\mathbf{ac}_1^*$ *and* $\mathbf{wac}_1^*$ *are equivalent, respectively, to* **Lin** *and* **SWLEM**.

Proof. Let α and β be sentences, and define

$$\theta(x, y) \equiv x = 0 \wedge [(y = 0 \wedge \alpha) \vee (y = 1 \wedge \beta)].$$

Then $\alpha \leftrightarrow \theta(0, 0)$ and $\beta \leftrightarrow \theta(0, 1)$, and so

$$\forall x \in 1[\theta(x, 0) \lor \theta(x, 1)] \leftrightarrow \theta(0, 0) \lor \theta(0, 1) \leftrightarrow \alpha \lor \beta.$$

Therefore

$$\exists f \in 2^1 [\forall x \in 1[\theta(x, 0) \lor \theta(x, 1)] \to \forall x \in 1\, \theta(x, fx)] \leftrightarrow \exists f \in 2^1[\alpha \lor \beta \to \theta(0, f0)]$$
$$\leftrightarrow [\alpha \lor \beta \to \theta(0, 0)] \lor [\alpha \lor \beta \to \theta(0, 1)]$$
$$\leftrightarrow [\alpha \lor \beta \to \alpha] \lor [\alpha \lor \beta \to \beta]$$
$$\leftrightarrow [\beta \to \alpha \lor \alpha \to \beta].$$

This yields $\mathbf{ac}_1^* \to \mathbf{Lin}$. For the converse, define $\alpha \equiv \theta(0,0)$ and $\beta \equiv \theta(0,1)$ and reverse the argument.

To establish the second stated equivalence, notice that, when $\theta(x,y)$ is defined as above, but with β replaced by $\neg \alpha$, it satisfies the provisions imposed in $\mathbf{wac}_1^*$. As above, that principle gives $(\neg \alpha \to \alpha) \lor (\alpha \to \neg \alpha)$, whence $\neg \alpha \lor \neg\neg \alpha$. So SWELM follows from $\mathbf{wac}_1^*$. Conversely, suppose that θ meets the condition imposed in $\mathbf{wac}_1^*$. Then from $\theta(0, 0) \to \neg\theta(0, 1)$ we deduce $\neg\neg\theta(0, 0) \to \neg\theta(0,1)$; now, assuming SWLEM, we have $\neg\theta(0, 0) \lor \neg\neg\theta(0, 0)$, whence $\neg\theta(0, 0) \lor \neg\theta(0, 1)$. Since $\neg\theta(0, 0) \to [\theta(0, 0) \to \theta(0, 1)]$ and $\neg\theta(0, 1) \to [\theta(0, 1) \to \theta(0, 0)]$ we deduce $[\theta(0, 0) \to \theta(0, 1)] \lor [\theta(0, 1) \to \theta(0, 0)]$. From the argument above it now follows that $\exists f \in 2^1 [\forall x \in 1[\varphi(x, 0) \lor \varphi(x, 1)] \to \forall x \in 1\, \varphi(x, fx)]$. Accordingly $\mathbf{wac}_1^*$ is a consequence of SWLEM. ∎

Chapter II

Natural Numbers and Finite Sets

THE NATURAL NUMBERS

The natural numbers can be defined in **IZ** and their usual properties proved.

Let us call a set A *inductive* if

$$0 \in A \land \forall x(x \in A \to x^+ \in A).$$

It follows from the axiom of infinity that there exists at least one inductive set A.

Define

$$\mathbb{N} = \bigcap \{X : X \subseteq A \land X \text{ is inductive}\}.$$

Then $\mathbb{N}$ is inductive and is clearly the *least* inductive set, that is, $\mathbb{N} \subseteq K$ for every inductive set K. The members of $\mathbb{N}$ are called *natural numbers*; thus 0, 1, 2 are natural numbers. We shall use letters $m, n, p, \ldots$ as variables ranging over $\mathbb{N}$.

Proposition 1.

(i) $m \in n \to m^+ \subseteq n$.
(ii) $n \notin n$
(iii) $m^+ = n^+ \to m = n$.

Proof. (i) Let $K = \{n : \forall m(m \in n \to m^+ \subseteq n)\}$. To prove (i) it suffices to show that K is inductive. Clearly $0 \in K$. Suppose now that $n \in K$. Then

(*) $\qquad\qquad\qquad m \in n \to m^+ \subseteq n.$

If $m \in n^+$ then $m \in n$ or $m = n$. In the first case $m^+ \subseteq n$ by (*) and in the second case $m^+ = n^+$. Thus in both cases $m^+ \subseteq n^+$ and so $n^+ \in K$. Hence K is inductive and **(i)** is proved.

(ii) It suffices to show that the set $L = \{n: n \notin n\}$ is inductive. Clearly $0 \in L$. If $n \in L$ and $n^+ \in n^+$, then $n^+ \in n$ or $n^+ = n$. In the first case, it follows from **(i)** that $n^{++} \subseteq n$, and since $n \in n^{++}$, that $n \in n$. The second case also implies $n \in n$. Thus in both cases $n \notin L$. So $n \in L$ and $n^+ \in n^+$ together lead to a contradiction, whence $n \in L \to n^+ \notin n^+$. i.e. $n \in L \to n^+ \in L$. So L is inductive and **(ii)** follows.

(iii) It follows from $m^+ = n^+$ that $m \in n^+$; thus $m \in n$ or $m = n$ and so by **(i)** $m \subseteq n$. Similarly $n \subseteq m$. ∎

We shall sometimes write $m < n$ for $m \in n$ and $m \leq n$ for $m \subseteq n$. It follows from Proposition 1 that $m^+ \leq n \leftrightarrow m < n$.

Proposition 2. *For arbitrary m, n, exactly one of the following holds:*

$$m \in n, \quad m = n, \quad n \in m.$$

Proof. Proposition 1 implies that any two of the above assertions are mutually contradictory. To prove that, for every m, n, one of these assertions holds we define

$$K(n) = \{m: m \in n \vee m = n \vee n \in m\}.$$

We need to show that $K(n) = \mathbb{N}$ for every n, and for this it suffices to show that $K(n)$ is inductive.

The set $K(0)$ is inductive, since $K(0) = \{0\} \cup \{m: 0 \in m\}$ and it is obvious that $0 \in m \to 0 \in m^+$.

Now suppose that $K(n)$ is inductive, i.e. $\mathbb{N} \subseteq K(n)$. We show that $K(n^+)$ is also inductive.

$0 \in K(n^+)$. From the fact that $K(0)$ is inductive, it follows that $n^+ \in \mathbb{N} \subseteq K(0)$, whence $n^+ \in 0$ or $n^+ = 0$ or $0 \in n^+$. The first two disjuncts are false, so $0 \in n^+$ and $0 \in K(n^+)$.

$\underline{m \in K(n^+) \to m^+ \in K(n^+)}.$ Suppose that $m \in K(n^+)$, that is, $m \in n^+$ or $m = n^+$ or $n^+ \in m$. In the second and third cases we obviously have $n^+ \in m^+$ and hence $m^+ \in K(n^+)$. In the first case either $m = n$ or $m \in n$. If $m = n$ then $m^+ = n^+$ so that $m^+ \in K(n^+)$. If $m \in n$, then $m \in K(n)$, and so, since $K(n)$ has been assumed inductive, $m^+ \in K(n)$. It follows that $m^+ \in n$ or $m^+ = n$ or $n \in m^+$. The third disjunct is false, since it implies (using Prop 1 (i)) that $n \in n$, contradicting Prop. 1(ii). Accordingly we have only the two possibilities $m^+ \in n$ or $m^+ = n$, both of which, since $n \subseteq n^+$, yield $m^+ \in K(n^+)$. The proof is complete. ∎

From Propositions 2 and 1(ii) we deduce immediately the

Corollary. $\mathbb{N}$ *is discrete.* ∎

MODELS OF PEANO'S AXIOMS

A *Peano structure* is a triple $\mathbf{A} = (A, s, o)$ where A is a set, $s: A \to A$, and $o \in A$.[12] A is called the *domain* of $\mathbf{A}$ and s the *successor operation* in $\mathbf{A}$. A *model of Peano's axioms* is a Peano structure $\mathbf{A}$ such that the following axioms are satisfied:

P1 $\forall p \in A . \ sp \neq o$
P2 $\forall p \in A \ \forall q \in A. \ sp = sq \to p = q.$
P3 $\forall X \subseteq A[[o \in X \wedge \forall p(p \in X \to sp \in X)] \to X = A]$

(P3) is the *Induction Principle* for $\mathbf{A}$: it is clearly equivalent to the scheme: for any formula $\varphi(x)$,

$$[\varphi(0) \wedge [\forall p \in A \ (\varphi(p) \to \varphi(sp))]] \to \forall p \in A \ \varphi(p)$$

---

[12] Here "o" (Greek omicron) is not to be confused with the set 0. It is just an arbitrary member of A.

A subset K of A satisfying $o \in K \wedge \forall x(x \in K \to sx \in K)$ is called an *inductive subset* of A: the Induction Principle asserts that the only inductive subset of A is A itself. Establishing that a subset K is inductive is called a *proof by induction*.

The following facts are easily established by induction:

In any model **A** of Peano's axioms,

$$sp \neq p$$
$$\forall p \in A[p = o \vee \exists q.\ p = sq].$$

If we define $s\colon \mathbb{N} \to \mathbb{N}$ by $sn = n^+$, then it follows from the fact that Prop.1 (iii) and the fact that $\mathbb{N}$ is the least inductive set that $(\mathbb{N}, s, 0)$ *is a model of Peano's axioms*.

DEFINITIONS BY RECURSION

Just as in classical set theory, in **IZ** any model of Peano's axioms admits functions defined by recursion. Let us say that a Peano structure $\mathbf{A} = (A, s, o)$ satisfies the *Simple Recursion Principle* if:

Given any set X, any element $a \in X$, and any function $e\colon X \to X$, there exists a unique function $f\colon A \to X$ such that

$$f(o) = a \quad \forall p \in A\ f(sp) = e(fp).$$

Proposition 3 *Any model of Peano's axioms satisfies the Simple Recursion Principle.*

Proof. Let $\mathbf{A} = (A, s, o)$ be a model of Peano's axioms. To simplify notation we shall use letters p, q to denote variables ranging over A.

Define

$$U = \{u \in \mathbf{P}(A \times X) : \langle o, a \rangle \in u \wedge \forall p \forall x (\langle p, x \rangle \in u \to \langle sp, e(x) \rangle \in u)\}.$$

Let $f = \bigcap U$. We claim that f satisfies the conditions of the proposition; its uniqueness is left as an exercise to the reader. To show that f satisfies the conditions of the proposition it clearly suffices to show that f is a map from A to X.

Clearly $f \subseteq A \times X$. Let $V = \{p : \exists x \langle p, x \rangle \in f\}$. Then $V \subseteq A$ and $o \in V$ since $\langle o, a \rangle \in f$. Moreover

$$p \in V \to \exists x \langle p, x \rangle \in f \to \exists x \langle sp, e(x) \rangle \in f \to sp \in V$$

So V is inductive, whence $V = A$, and f is defined on A.

It remains to show that f is single-valued. To this end define

$$K = \{p : `\forall x \forall y (\langle p, x \rangle \in f \wedge \langle p, y \rangle \in f \to x = y)\}.$$

We need to show that K is inductive.

First, we show that $o \in K$. Define

$$f^* = \{\langle p, x \rangle \in f : p = o \to x = a\}.$$

It is easily verified that $f^* \in U$, whence $f = f^*$. Therefore

$$\langle o, x \rangle \in f \to \langle o, x \rangle \in f^* \to x = a,$$

so that
$$\forall x (\langle o, x \rangle \in f \to x = a),$$

from which it follows immediately that $o \in K$.

Finally we need to show that $p \in K \to sp \in K$. To do this we first establish the auxiliary result

(1) $\qquad (p \in K \wedge \langle p, x \rangle \in f \wedge \langle sp, y \rangle \in f) \to y = e(x).$

Given $p \in K$ satisfying $\langle p, x \rangle \in f \wedge \langle sp, y \rangle \in f$, define

$$f_{px} = \{\langle q, y \rangle \in f : q = sp \to y = e(x)\}.$$

We claim that $f_{px} \in U$. First, from **P1** it follows that $\langle 0, a \rangle \in f_{px}$. Now, if $\langle q, y \rangle \in f_{px}$, then $\langle q, y \rangle \in f$ and $q = sp \to y = e(x)$. We need to show that $\langle sq, e(y) \rangle \in f_{px}$ and for this it suffices to show that

(2) $$\langle sq, e(y) \rangle \in f$$

and

(3) $$sp = sq \to e(y) = e(x).$$

Assertion (2) follows from the assumption that $\langle q, y \rangle \in f$. As for (3), if $sp = sq$, then, by **P2**, $p = q$, so from $\langle q, y \rangle \in f$ it follows that $\langle p, y \rangle \in f$. But we are assuming that $p \in K$ and $\langle p, x \rangle \in f$, so we conclude from the defining property of K that $x = y$. Hence, certainly, $e(x) = e(y)$, proving (2).

We conclude that $f_{px} \in U$. From this it follows that $f \subseteq f_{px}$, so that

$$(p \in K \wedge \langle p, x \rangle \in f \wedge \langle sp, y \rangle \in f) \to f_{px} \in U$$
$$\to f \subseteq f_{px},$$

whence

$$(p \in K \wedge \langle p, x \rangle \in f \wedge \langle sp, y \rangle \in f) \to \langle sp, y \rangle \in f_{px}$$
$$\to y = e(x).$$

Thus (1) is proved.

Now, we know that $V = A$, i.e. $\forall p \exists z. \langle p, x \rangle \in f$. it follows from this and (1) that

$$(p \in K \wedge \langle sp, y \rangle \in f \wedge \langle sp, z \rangle \in f)$$
$$\to \exists x. \langle p, x \rangle \in f \wedge p \in K \wedge \langle sp, y \rangle \in f \wedge \langle sp, z \rangle \in f$$
$$\to \exists x (p \in K \wedge \langle p, x \rangle \in f \wedge \langle sp, y \rangle \in f \wedge \langle sp, z \rangle \in f)$$
$$\to \exists x (y = e(x) \wedge z = e(x))$$
$$\to y = z.$$

Therefore $p \in K \to sp \in K$, and it follows that K is inductive, so that $K = A$ and f is single-valued. This completes the proof of Proposition 3. ∎

An *isomorphism* between two Peano structures $\mathbf{A} = (A, s, o)$ and $\mathbf{A'} = (A', s', o')$ is a bijection $f: A \to A'$ such that $f(o) = o'$ and for all $p \in A$, $f(s(p)) = s'(f(p))$. Two Peano structures are *isomorphic* if there is an isomorphism between them.

Corollary 1. *Any two models of Peano's axioms are isomorphic.*

Proof. Given two Peano structures $\mathbf{A} = (A, s, o)$ and $\mathbf{A'} = (A', s', o')$, by Prop. 3 they both satisfy the Simple Recursion Principle, so that there are maps $f: A \to A'$ and $g: A' \to A$ such that

$$f(o) = o' \ \& \ \text{for all } p \in A, f(s(p)) = s'(f(p)).$$
$$g(o') = o \ \& \ \text{for all } q \in A', g(s'(q)) = s(g(q)).$$

We claim that f is an isomorphism between $\mathbf{A}$ and $\mathbf{A'}$. For this to be the case it suffices to show that g is an inverse to f, i.e. $g \circ f = 1_A$ and $f \circ g = 1_{A'}$. To prove the first assertion it is enough to show that the set $K - \{p \in A: g(f(p)) = p \}$ is inductive, and this is a straightforward consequence of the definitions of f and g. The proof of the second assertion is similar. ∎

Corollary 2. *The domain of any Peano structure is discrete.*

Proof. By the Corollary to Prop. 3, $\mathbb{N}$ is discrete, and is the domain of a model of Peano's axioms. If A is the domain of a model of Peano's axioms, it is, by Corollary 1, bijective with $\mathbb{N}$, and it follows easily from this and the discreteness of $\mathbb{N}$ that A is discrete. ∎

Let us say that a Peano structure **A** = (A, s, o) satisfies the *Extended Recursion Principle* if:

given any set X, any element $a \in X$, and any function $e: X \times A \to X$, there exists a unique function $f: A \to X$ such that

$$f(o) = a \;\land\; \forall p \in A.\; f(sp) = e(\langle fp, p \rangle).$$

Proposition 4. *Any Peano structure that satisfies the Simple Recursion Principle also satisfies the Extended Recursion Principle.*

Proof. Let **A** = (A, s, o) be a Peano structure satisfying the Simple Recursion Principle. We use p as a variable ranging over A. Given a set X, an element $a \in X$, and a function $e: X \times A \to X$, let $Y = X \times A$ and let $h: Y \to Y$ be the function given by

$$h = \langle x, p \rangle \mapsto \langle e(\langle x, p \rangle), p \rangle$$

Applying the Simple Recursion Principle to h, Y and $\langle a, o \rangle$ yields a unique $k: A \to Y$ such that

$$k(o) = \langle a, o \rangle \land \forall p.\; k(sp) = h(k(p)).$$

It is now easily checked that $f = \pi_1 \circ k : A \to X$ is the unique map such that

$$f(o) = a \;\land\; \forall p \in A.\; f(sp) = e(\langle fp, p \rangle).$$

The Extended Recursion Principle follows. ∎

We use Proposition 4 to prove the converse of Proposition 3, namely

Proposition 5. *Any Peano structure that satisfies the Simple Recursion Principle is a model of Peano's axioms.*

Proof. Let **A** = (A, s, o) be a Peano structure satisfying the Simple Recursion Principle. Then by Prop. 4 **A** also satisfies the Extended Recursion Principle. We need to verify that **A** satisfies **P1, P2** and **P3**. Again we use p, q as variables ranging over A.

P1: $\forall p . sp \neq o$. Recalling that $2 = \{0, 1\}$, define $g: 2 \to 2$ by $g = \{\langle 0,1 \rangle, \langle 1,1 \rangle\}$. Using the Simple Recursion Principle, there is $f: P \to 2$ such that

$$f(o) = 0 \wedge \forall p.\ f(sp) = g(fp).$$

Then

$$sp = o \to 0 = f(o) = f(sp) = g(fp) = 1;$$

since $0 \neq 1$, **P1** follows.

P2: $\forall p \forall q.\ sp = sq \to p = q$. Consider $\pi_2 : A \times A \to A$. By the Extended Recursion Principle there is a map $f: A \to A$ such that

$$f(o) = o \wedge \forall p.\ f(p) = \pi_2(\langle fp, p \rangle) = p.$$

Then

$$sp = sq \to q = f(sq) = f(sp) = p$$

and **P2** follows.

P3: the Induction Principle. Suppose $K \subseteq A$ is inductive. Then $p \mapsto sp: K \to K$ and so the Simple Recursion Principle furnishes a map $f: A \to K$ such that

$$f(o) = o \wedge \forall p.\ f(sp) = s(fp).$$

Writing j for the insertion map of K into A, we get

(*) $\qquad (j \circ f)(o) = o \wedge \forall p.\ (j \circ f)(sp) = s(j \circ f)(p).$

But the identity map $1_A: A \to A$ also satisfies (*), so from the uniqueness condition in the Simple Recursion Principle we infer that $j \circ f = 1_A$. It follows easily from this that $K = A$, and the Induction Principle follows. ∎

A set A is *Dedekind infinite* if there exists a monic map $f: A \to A$ and an element $a \in A$ such that $a \notin \text{ran}(f)$. We next show that each Dedekind infinite set gives rise to a model of Peano's axioms.

Proposition 6. *Each Dedekind infinite set contains the domain of a model of Peano's axioms.*

Proof. Let A be Dedekind infinite, $f: A \to A$ monic, and $a \in A$ such that $a \notin \text{ran}(f)$. Define

$$U = \bigcap \{u : u \subseteq A \wedge a \in U \wedge \forall x \in u. f(x) \in u\}.$$

It is then easily shown that (U, f, a) is a model of Peano's axioms. ∎

Corollary. *A set A is Dedekind infinite if and only if there exists an injection $\mathbb{N} \to A$.*

FINITE SETS

There are a number of possible definitions of the concept of *finite set* in **IZ**. To introduce (some of) these, it will be convenient to fix a set E. By an *E-family* or *E-singleton* we shall mean "set of subsets of E", or "singleton of E", respectively. For a subset X of E we define

$K(X) \equiv X$ is in every E-family containing $\emptyset$, all E-singletons, and closed under unions of pairs of its members. If $K(X)$ holds, we shall say that X is a *Kuratowski finite* subset of E.

$L(X) \equiv X$ is in every E-family containing $\emptyset$ and closed under unions with E-singletons. If $L(X)$ holds, we shall say that X is a *finite* subset of E.

$M(X) \equiv X$ is in every E-family $\mathscr{F}$ containing $\emptyset$ and closed under unions with disjoint E-singletons, that is, if $\forall X \in \mathscr{F} \, \forall x \in E - X (X \cup \{x\} \in \mathscr{F})$. If $M(X)$ holds, we shall say that X is a *strictly finite* subset of E.

We shall also write $D(X)$ for "X is discrete".

Lemma 1. $\forall X \subseteq E[M(X) \to L(X)]$.

Proof. Obvious. ∎

Lemma 2. $\forall X \subseteq E\ [K(X) \leftrightarrow L(X)]$.

Proof. Clearly $L(X) \to K(X)$. To prove the converse, it suffices to show that the family $\mathscr{L} = \{X \subseteq E: L(X)\}$ is closed under unions of pairs. To this end let $\Phi(U)$ be the property $\forall X \in \mathscr{L}.\ U \cup X \in \mathscr{L}$. It suffices to show $\forall U[L(U) \to \Phi(U)]$. Clearly $\Phi(\emptyset)$. Assuming $\Phi(U)$ and $X \in \mathscr{L}$ we have $U \cup X \in \mathscr{L}$ and so $U \cup X \cup \{x\} \in \mathscr{L}$ for arbitrary x, whence $\Phi(U \cup \{x\})$. Hence $\forall U[L(U) \to \Phi(U)]$ and the result follows. ∎

Lemma 3. $\forall X[M(X) \to D(X)]$.

Proof. Obviously $D(\emptyset)$. If $D(X)$ and $x \notin X$, clearly $D(X \cup \{x\})$. The result follows. ∎

Lemma 4. $\forall X \subseteq E\ [M(X) \to \forall a[D(X \cup \{a\}) \to (a \in X \vee a \notin X)]]$.

Proof. Write $\Phi(X)$ for the condition following the first implication. Clearly $\Phi(\emptyset)$. Suppose that $\Phi(X)$ and $x \notin X$. If $D(X \cup \{x\} \cup \{a\})$, then $D(X \cup \{a\})$, so, since $\Phi(X)$, either $a \in X \vee a \notin X$. Since $D(X \cup \{x\} \cup \{a\})$, it follows that $a = x \vee a \neq x$. Hence

$$(a \in X \wedge a = x) \vee (a \notin X \wedge a = x) \vee (a \in X \wedge a \neq x) \vee (a \notin X \wedge a \neq x),$$

The first three disjuncts each imply $a \in X \cup \{x\}$, and the last disjunct means $a \notin X \cup \{x\}$.

Accordingly $a \in X \cup \{x\} \vee a \notin X \cup \{x\}$. We conclude that $\Phi(X \cup \{x\})$ and the result follows. ∎

Lemma 5. $\forall X \subseteq E\ [L(X) \wedge D(X) \to M(X)]$.

Proof. We need to show $\forall X[L(X) \to \Phi(X)]$, where $\Phi(X)$ is $D(X) \to M(X)$. Clearly $\Phi(\emptyset)$. Assume $\Phi(X)$ and $D(X \cup \{a\})$. Then $D(X)$, so, since $\Phi(X)$, it follows that

$M(X)$. Since $D(X \cup \{a\})$, Lemma 4 gives $a \in X \vee a \notin X$. In either case we deduce that $M(X \cup \{a\})$. Hence $\Phi(X \cup \{a\})$, and the result follows. ∎

From these lemmas we immediately infer

Proposition 7. *For any set E, the families of strictly finite, discrete finite, and discrete Kuratowski finite subsets coincide.* ∎

We can now define a set E to be *strictly finite, finite,* or *Kuratowski finite* if it is, respectively, a strictly finite, finite, or Kuratowski finite subset of itself.

Proposition 8. *A set is strictly finite if and only if it is bijective with a natural number.*

Proof. Suppose that E is strictly finite, and for $X \subseteq E$ let $\Phi(X)$ be the property X is bijective with a natural number. We need to show that $\Phi(E)$, and for this it suffices to show that, for all $X \subseteq E$, $M(X) \to \Phi(X)$. Clearly we have $\Phi(\emptyset)$. If $\Phi(X)$ and $x \in E - X$, let $f: n \to X$ be a bijection between some natural number n and X. It is easily checked that the set $g = f \cup \{<n,x>\}$ is a bijection between n^+ and $X \cup \{x\}$. Hence $\Phi(X \cup \{x\})$ and the result follows.

Conversely, suppose that E is bijective with a natural number, i.e. $\Phi(E)$. We want to show that E is strictly finite, and for this it suffices to show that the subset K of $\mathbb{N}$ given by

$$K = \{n \in \mathbb{N} : \forall X \in \mathbf{P}E (X \approx n \to X \text{ strictly finite}\}$$

is inductive. Clearly $0 \in K$. Now suppose $n \in K$, and $X \approx n^+$. Let f be a bijection between n^+ and X, let $a = f(n)$ and let $X' = X - \{a\}$. Then the restriction $f|n$ is a bijection between n and X' and so, since $n \in K$, it follows that X' is strictly finite. But then, since $a \notin X'$, it follows that $X = X' \cup \{a\}$ is also strictly finite. Hence $n^+ \in K$, so that K is inductive. This completes the proof. ∎

Finally, a set E is *Dedekind finite* if it is not Dedekind infinite, i.e. if there does not exist a monic $f: E \to E$ and an element $a \in E$ such that $a \notin \text{ran}(f)$.

Proposition 9. *Every strictly finite set is Dedekind finite.*

Proof. To prove this, it suffices to show that $\emptyset$ is Dedekind finite and

(*) $\qquad\qquad X$ Dedekind finite & $a \notin X \to X \cup \{a\}$ Dedekind finite.

It is obvious that $\emptyset$ is Dedekind finite. To prove (*), suppose that $a \notin X$ and $X \cup \{a\}$ is Dedekind infinite. We show that X is Dedekind infinite. Since $X \cup \{a\}$ is Dedekind infinite, there is a monic $f: X \cup \{a\} \to X \cup \{a\}$ and $b \in X \cup \{a\}$ such that $b \notin \text{ran}(f)$. There are two cases: $b = a$ or $b \in X$. In the first case, $f: X \cup \{a\} \to X$; the restriction $f|X : X \to X$ is then monic and $f(a) \in X - \text{ran}(f|X)$, so that X is Dedekind infinite. The second case, $b \in X$, splits into two subcases: $f(a) = a$ or $f(a) \in X$. In the first subcase, $f|X : X \to X$ is monic and $b \in X - \text{ran}(f|X)$, so that again X is Dedekind-infinite. In the second subcase, $f(a) \in X$, there is a unique $x_0 \in X$ for which $f(x_0) = a$, so that $x_0 \notin f^{-1}[X]$ and it is then easily shown that $X = f^{-1}[X] \cup \{x_0\}$. Now define

$$g = f \mid f^{-1}[X] \cup \{\langle x_0, f(a)\rangle\}$$

It is then easily shown that g is a monic map from X to X and $b \notin \text{ran}(g)$. Thus X is Dedekind infinite and the result is proved. ∎

We conclude this section with

Proposition 10. Ω *is Dedekind finite.*

This is an immediate consequence of

Proposition 11. *If $f: \Omega \to \Omega$ is monic, then $f^2 = 1_\Omega$, so that f is also epi..*

Proof. In the proof we shall use the easily established fact that *for* $\omega, \upsilon \in \Omega$,

(1) $\qquad\qquad \omega = \upsilon \leftrightarrow (\omega = 1 \leftrightarrow \upsilon = 1)$.

We first prove

(2) $$f(\omega) = 1 \to f^2(\omega) = \omega.$$

Assume $f(\omega) = 1$. We show that $f^2(\omega) = 1 \leftrightarrow \omega = 1$, from which (2) then follows by (1).

First, we have
$$\omega = 1 \to \omega = f(\omega) \to f^2(\omega) = f(\omega) = 1.$$

Conversely
$$f^2(\omega) = 1 \to f^2(\omega) = 1 = f(\omega) \to \omega = f(\omega) = 1,$$

as required.

Finally we use (2) to prove

(3) $$f^3(\omega) = f(\omega),$$

from which we infer $f^2(\omega) = \omega$, so that $f^2 = 1_\Omega$.

To prove (3), by (1) it suffices to show that

(4) $$f^3(\omega) = 1 \leftrightarrow f(\omega) = 1.$$

If $f(\omega) = 1$, it follows from (2) that $f^2(\omega) = \omega$, whence $f^3(\omega) = f(\omega) = 1$. Conversely, if $f^3(\omega) = 1$, then $f(f^2(\omega)) = 1$, so by (2) $f^4(\omega) = f^2(f^2(\omega)) = f^2(\omega)$. It follows that $1 = f^3(\omega) = f(\omega)$. This proves (4), and the Proposition. ∎

FREGE'S CONSTRUCTION OF THE NATURAL NUMBERS

By *Frege's Theorem* is meant the result, implicit in Frege's *Grundlagen*, that, for any set E, if there exists a map v from PE to E satisfying the condition

$$\forall X \forall Y [\, v(X) = v(Y) \leftrightarrow X \approx Y\,],$$

then E has a subset which is the domain of a model of Peano's axioms. We are going to show that a strengthened version of this result can be proved in **IZ**.

Let us call a family of subsets of a set E *strictly inductive* if it contains $\emptyset$ and is closed under unions with disjoint E-singletons. We define a *Frege structure* to be a pair (E, v) with v a map to E whose domain $\mathrm{dom}(v)$ is a strictly inductive family of subsets of E such that

$$\forall X \in \mathrm{dom}(v) \forall Y \in \mathrm{dom}(v) \ [v(X) = v(Y) \leftrightarrow X \approx Y].$$

A Frege structure (E, v) is *strict* if $\mathrm{dom}(v)$ is the family of strictly finite subsets of E.

We now prove

Frege's Theorem. *Let (E, v) be a Frege structure. Then we can define a subset N of E which is the domain of a model of Peano's axioms.*

Thus suppose given a Frege structure (E, v). The proof of Frege's Theorem breaks down into a sequence of lemmas.

For $X \in \mathrm{dom}(v)$ write $X^\dagger$ for $X \cup \{v(X)\}$. Call a property Φ defined on the members of $\mathrm{dom}(v)$ *v-inductive* if $\Phi(\emptyset)$ and, for any $X \in \mathrm{dom}(v)$, if $\Phi(X)$ and $v(X) \notin X$, then $\Phi(X^\dagger)$. Call a subfamily $\mathscr{A}$ of $\mathrm{dom}(v)$ *v-inductive* if the property of being a member of $\mathscr{A}$ is v-inductive. Then $\mathrm{dom}(v)$ is v-inductive, as is the intersection $\mathscr{N}$ of the collection of all v-inductive families. From the fact that $\mathscr{N}$ is the least v-inductive family we infer immediately the

Principle of v-Induction for $\mathscr{N}$. *For any property Φ defined on the members of $\mathscr{N}$, if Φ is v-inductive, then every member of $\mathscr{N}$ has Φ.*

Lemma 1. *For any $X \in \mathscr{N}$,*

$$X = \emptyset \text{ or } X = Y^\dagger \text{ for some } Y \in \mathscr{N} \text{ such that } v(Y) \notin Y.$$

Proof. Write $\Phi(X)$ for this assertion. To establish the claim it is enough, by the Principle of v- Induction, to show that Φ is v-inductive. Clearly $\Phi(\emptyset)$. If $\Phi(X)$ and $v(X) \notin X$, then evidently $\Phi(X^+)$. So Φ is v-inductive. ∎

Lemma 2. *For any $X \in \mathcal{N}$ and any $x \in X$,*

$$\text{there is } Y \in \mathcal{N} \text{ such that } Y \subseteq X \text{ and } x = v(Y).$$

Proof. Writing $\Phi(X)$ for this assertion, it suffices to show that Φ is v-inductive. Clearly $\Phi(\emptyset)$. Now assume $\Phi(X)$ and $x \in X^+$. Then either $x \in X$, in which case, since $\Phi(X)$ has been assumed, there is $Y \in \mathcal{N}$ for which $x = v(Y)$ and $Y \subseteq X$, a fortiori $Y \subseteq X^+$. Or $x = v(X)$, yielding the same conclusion with $Y = X$. So we obtain $\Phi(X^+)$, Φ is v- inductive, and the Lemma follows. ∎

Lemma 3. *If $X, Y \subseteq E$, $x \in E - X$, $y \in E - Y$, and $X \cup \{x\} \approx Y \cup \{y\}$, then $X \approx Y$.*

Proof. Assume the premises and let f be a bijection between $X \cup \{x\}$ and $Y \cup \{y\}$. We produce a bijection f' between X and Y. Let y' be the unique element of $Y \cup \{y\}$ for which $\langle x, y'\rangle \in f$. Then either $y' = y$, in which case we take $f' = f \mid X$, or $y' \in Y$, in which case the unique element $x' \in X \cup \{x\}$ for which $\langle x', y\rangle \in f$ satisfies $x' \in X$. (For if $x' = x$ then $\langle x, y\rangle \in f$, in which case $y' = y \notin Y$.) So in this case we define

$$f' = [f \cap (X \times Y)] \cup \{\langle x', y'\rangle\}.$$

In either case it is easily checked that f' is a bijection between X and Y. This proves the Lemma. ∎

Lemma 4. *For all X, Y in $\mathcal{N}$,*

$$v(X) = v(Y) \to X = Y.$$

Proof. Write $\Phi(X)$ for the assertion $X \in \mathcal{N}$ and $\forall Y \in \mathcal{N}[v(X) = v(Y) \to X = Y]$. It suffices to show that Φ is v-inductive. $\Phi(\emptyset)$ holds because $v(\emptyset) = v(Y) \to Y \approx \emptyset \to \emptyset = Y$. Now assume that $\Phi(X)$ and $v(X) \notin X$; we derive $\Phi(X^+)$. Suppose that $Y \in N$ and $v(X^+) = v(Y)$. Then $X^+ \approx Y$, and so in particular $Y \neq \emptyset$. By Lemma 1, there is $Z \in \mathcal{N}$ for which $v(Z) \notin Z$ and $Y = Z^\dagger$, so that $X^\dagger \approx Z^\dagger$. We deduce, using Lemma 3, that $X \approx Z$, so, since we have assumed $\Phi(X)$, $X = Z$. Hence $X^\dagger = Z^\dagger = Y$, and $\Phi(X^\dagger)$ follows. So Φ is v-inductive and the Lemma proved. ∎

Lemma 5. *For any* $X \in \mathcal{N}$,

$$v(X) \notin X.$$

Proof. It suffices to show that the property $v(X) \notin X$ is v-inductive. Obviously $\emptyset$ has this property. Supposing that $X \in \mathcal{N}$, $v(X) \notin X$ but $v(X^\dagger) \in X^\dagger$, we have either $v(X^\dagger) = v(X)$ or $v(X^\dagger) \in X$. In the former case $X = X^\dagger$ by Lemma 4, so that $v(X) \in X$, a contradiction. In the latter case, by Lemma 2, there is $Y \in \mathcal{N}$ such that $Y \subseteq X$ and $v(X^\dagger) = v(Y)$. Lemma 4 now applies to yield $X^\dagger = Y \subseteq X$, so again $v(X) \in X$, a contradiction. Therefore $v(X) \notin X \to v(X^\dagger) \notin X^\dagger$, and the Lemma follows. ∎

Notice that it follows immediately from Lemma 5 that $\mathcal{N}$ is closed under †, that is, $X \in \mathcal{N} \to X^\dagger \in \mathcal{N}$.

Now define $o = v(\emptyset)$, $N = \{v(X): X \in \mathcal{N}\}$, and $s: N \to N$ by $s(v(X)) = v(X^\dagger)$ for $X \in \mathcal{N}$. Then s is well defined and monic on N. (For if $v(X) = v(Y)$, then, by Lemma 4, $X = Y$, and so $s(v(X)) = v(X^\dagger) = v(Y^\dagger) = s(v(Y))$. Conversely, if $s(v(X)) = s(v(Y))$, then $v(X^\dagger) = v(Y^\dagger)$, so that, by Lemma 4, $X^+ \approx Y^+$. Lemmas 3 and 5 now imply $X \approx Y$, whence $v(X) = v(Y)$.) Clearly, also, $o \neq sx$ for any $x \in N$. The fact that the structure $(N, s, 0)$ satisfies the Induction Principle follows immediately from the Principle of v-induction for $\mathcal{N}$. Accordingly (N, s, o) is a model of Peano's axioms, as required.

The proof of Frege's Theorem is complete.

We next establish a converse to Frege's Theorem, namely, that any set containing the domain of a model of Peano's axioms determines a map which turns the set into a strict Frege structure: And finally, we show that the procedures leading from strict Frege structures to models of Peano's axioms and vice-versa are *mutually inverse*. It follows that the postulation of a (strict) Frege structure is constructively equivalent to the postulation of a model of Peano's axioms.

Let $\mathbf{A} = (A, s, o)$ be a model of Peano's axioms; we use letters p, q, r as variables ranging over A. Using the Simple Recursion Principle, define $k: A \to \mathbf{P}A$ to satisfy the equations

$$k(o) = \emptyset \qquad k(sp) = k(p) \cup \{p\}.$$

Lemma 6. (i) $p \in k(q) \to k(p) \subseteq k(q)$.
(ii) $p \notin k(p)$.

Proof. (i) Let $K = \{q: \forall p[p \in k(q) \to k(p) \subseteq k(q)]\}$. Obviously $o \in K$. If $q \in K$, then $p \in k(q) \to k(p) \subseteq k(q)$, so that

$$p \in k(sq) = k(q) \cup \{q\} \to p \in k(q) \lor p = q. \to k(p) \subseteq k(sq).$$

Thus K is inductive and **(i)** follows.

(ii). Let $K = \{p: p \notin k(p)\}$. Obviously $o \in K$. Suppose that $p \in K$. So if $sp \in k(sp)$, then $sp = p$ or $sp \in k(p)$. The first case is impossible and the second case, using (i), yields $k(p) \cup \{p\} = k(sp) \subseteq k(p)$ whence $p \in k(p)$, contradicting $p \in K$. Hence $sp \notin k(sp)$, i.e. $sp \in K$. Hence K is inductive and **(ii)** follows. ∎

Lemma 7. *For all p, q, $k(p) \approx k(q) \leftrightarrow p = q$.*

Proof. Write $\Phi(p)$ for $\forall q[k(p) \approx k(q) \leftrightarrow p = q]$. Then clearly $\Phi(o)$. If $\Phi(p)$ and $k(q) \approx k(sp) = k(p) \cup \{p\}$, then $q \neq o$ so that $q = sr$ for some r. Hence

$$k(r) \cup \{r\} = k(sr) = k(q) \approx k(sp) = k(p) \cup \{p\}.$$

Since, by (i) of Lemma 7, $r \notin k(r)$ and $p \notin k(p)$, Lemma 3 implies that $k(r) \approx k(p)$, so, since $\Phi(p)$, $r = p$ and $q = sr = sp$. Hence $\Phi(sp)$, and the result follows by induction. ∎

Now suppose that E is a set such that $A \subseteq E$. Define

$$v = \{<X,p> \in \mathbf{P}E \times A : X \approx k(p)\}.$$

Lemma 8. $\mathrm{dom}(v)$ *is the family of strictly finite subsets of* E.

Proof. We need to show that $\mathrm{dom}(v)$ is the least family of subsets of E which contains $\emptyset$ and is closed under unions with disjoint E-singletons: let us again call such a family *strictly inductive*. First, $\mathrm{dom}(v)$ clearly contains $\emptyset$. If $X \in \mathrm{dom}(v)$, then $X \approx k(p)$ for some p. If $x \notin X$, then $X \cup \{x\} \approx k(p) \cup \{p\} = k(sp)$, whence $X \cup \{x\} \in \mathrm{dom}(v)$. So $\mathrm{dom}(v)$ is strictly inductive. And $\mathrm{dom}(v)$ is the least strictly inductive family. For suppose that $\mathcal{F}$ is any strictly inductive family. For each p let $\mathcal{H}_p = \{X : X \approx k(p)\}$. We claim that $\mathcal{H}_p \subseteq \mathcal{F}$ for all p. For obviously $\mathcal{H}_o = \{\emptyset\} \subseteq F$. Now suppose that $\mathcal{H}_p \subseteq \mathcal{F}$. If $X \approx k(sp)$, then $X \approx k(p) \cup \{p\}$, so for some $x \in X$ (which may be taken to be the image of p under a bijection between $k(p) \cup \{p\}$ and X), we have $X - \{x\} \approx k(p)$. It follows that $X - \{x\} \in \mathcal{H}_n \subseteq \mathcal{F}$, and so $X = (X - \{x\}) \cup \{x\} \in \mathcal{F}$. The claim now follows by induction; accordingly $\mathrm{dom}(v)$, as the union of all the $\mathcal{H}_n$, is included in $\mathcal{F}$. Therefore $\mathrm{dom}(v)$ is the least inductive family and the Lemma is proved. ∎

Lemma 9. . v *is a function and* $X \approx k(v(X))$ *for all* $X \in \mathrm{dom}(v)$.

Proof. Suppose that $<X, p> \in v$ and $<X, q> \in v$. Then $X \approx k(p)$ and $X \approx k(q)$ whence $k(p) \approx k(q)$ and so $p = q$ by Lemma 7. The remaining claim is obvious. ∎

Lemma 10. *For all* $X, Y \in \mathrm{dom}(v), X \approx Y \leftrightarrow v(X) = v(Y)$.

Proof. We have, using the previous Lemma, $v(X) = v(Y) \leftrightarrow k(v(X)) \approx k(v(Y)) \leftrightarrow X \approx Y$. ∎

Lemmas 8 and 10 establish

Proposition 12. *(E, v) is a strict Frege structure.* ∎

(E,v) is called the strict Frege structure *associated* with E and the model **A** of Peano's axioms.

Finally, we show that the processes of deriving models of Peano's axioms from strict Frege structures and *vice-versa* are mutually inverse.

Suppose that we are given a strict Frege structure (E, μ). Recall that the associated model (A, s, o) of Peano's axioms is obtained in the following way. First, the family $\mathcal{N}$ is defined as the least subfamily of $\text{dom}(\mu)$ containing $\varnothing$ and such that, if $X \in \mathcal{N}$ and $\mu(X) \notin X$, then $X \cup \{\mu(X)\} \in \mathcal{N}$: it having been shown that $\mu(X) \notin X$ for all $X \in \mathcal{N}$. The associated model (A, s, o) of Peano's axioms was then defined by $N = \{\mu(X): X \in \mathcal{N}\}$, $s(\mu(X)) = \mu(X \cup \mu(X)\})$, and $o = \mu(\varnothing)$.

We observe that since (E, μ) is strict, for any $X \in \text{dom}(\mu)$ there is a (unique) $X^* \in \mathbf{N}$ for which $X \approx X^*$, and so $\mu(X) = \mu(X^*)$. To prove this, it suffices to show that the set of $X \in \text{dom}(\mu)$ with this property contains $\varnothing$ and is closed under unions with disjoint singletons. The first claim is obvious. If $X \in \text{dom}(\mu)$, $x \notin X$, and $X \approx X^*$ with $X^* \in \mathcal{N}$, then

$$X \cup \{x\} \approx X^* \cup \{\mu(X^*)\} \in \mathcal{N},$$

since, as observed above, $\mu(X^*) \notin X^*$. This establishes the second claim, and the observation.

Now let (E, v) be the strict Frege structure associated with the model (A, s, o) of Peano's axioms in turn associated with (E, μ). We claim that $\mu = v$. To prove this it suffices to show that

(*) $\qquad\qquad X \approx k(\mu(X))$ for all $X \in \mathcal{N}$,

where $\mathcal{N}$ is defined as above. For then, by Lemma 9, we will have $k(v(X)) \approx X \approx k(\mu(X))$ and so $\mu(X) = v(X)$ by Lemma 7. This last equality for all

$X \in \mathcal{N}$ in turn yields $\mu(Y) = v(Y)$ for all $Y \in \text{dom}(\mu) = \text{dom}(v)$. For, by our observation above, $\mu(Y) = \mu(Y^*) = v(Y^*) = v(Y)$.

So it only remains to prove (*). It is clearly satisfied by $\varnothing$. If $X \approx k(\mu(X))$ with $X \in \mathcal{N}$, then, since $\mu(X) \notin X$,

$$X \cup \{\mu(X)\} \approx k(\mu(X)) \cup \{\mu(X)\} = k(s\mu(X)).$$

(*) now follows from the definition of $\mathcal{N}$. So our claim that $\mu = v$ is established.

Conversely, suppose we are given a set E and a model (A, s, o) of Peano's axioms with $A \subseteq E$. Let (E, v) be the associated strict Frege structure. We note first that, for any $p \in A$, we have $v(k(p)) = p$. For by Lemma 9, $k(p) \approx k(v(k(p)))$, so that, by Lemma 7, $p = k(v(p))$. Now let (A^*, s^*, o^*) be the model of Peano's axioms associated with the Frege structure (E, v). We claim that (A, s, o) and (A^*, s^*, o^*) are identical.

First, $A^* = \{v(X): X \in \mathcal{N}^*\}$, where $\mathcal{N}^*$ is the least subfamily of $\text{dom}(v)$ containing $\varnothing$ and such that $X \in \mathcal{N}^*$ and $v(X) \notin X$ implies $X \cup \{v(X)\} \in \mathcal{N}^*$. Using the fact that $v(k(p)) = p$ for all $p \in A$, it is easily shown that $\mathcal{N}^* = \{k(p): p \in A\}$. Thus $A^* = \{v(X): X \in \mathcal{N}^*\} = \{v(kp)): p \in A\} = \{p: p \in A\} = A$. Finally $o^* = v(o) = v(g(o)) = o$ and

$$s^*(p) = s^*(v(k(p))) = v(k(p) \cup \{v(k(p))\}) = v(k(p) \cup \{p\}) = v(k(sp)) = sp,$$

so that $s^* = s$ by induction.

Thus we have established that the two processes are mutually inverse.

Chapter III

The Real Numbers

We turn now to the construction of the real numbers in **IZ**. This is done in essentially the classical manner: first, the (positive and negative) integers are constructed, next, the rationals, and then the (ordered set of) reals obtained as Dedekind cuts or Cauchy sequences. In the classical context it is well known that these methods of constructing the reals lead to isomorphic results. This is not necessarily the case in **IZ**. Nor is it necessarily the case that the reals are (conditionally) order-complete, or even discrete.

The set $\mathbb{Z}$ of positive and negative integers is constructed within **IZ** in the usual way[13]. It is shown in the standard wat that $\mathbb{Z}$ may be turned into an ordered ring $\langle \mathbb{Z}, +, \cdot, < \rangle$.

The customary procedure for obtaining the rational field as the ordered field of quotients of $\mathbb{Z}$ [14] now yields, in **IZ**, the *ordered field of rationals* $\langle \mathbb{Q}, +, \cdot, < \rangle$. Note that $\mathbb{Q}$, like $\mathbb{N}$ and $\mathbb{Z}$, is discrete. We shall use letters p, q as variables ranging over $\mathbb{Q}$.

Now we can define the Dedekind real numbers as "cuts" in $\mathbb{Q}$ Thus a *Dedekind real number* is a pair <**L**, **R**> of inhabited subsets **L**, **R** $\subseteq \mathbb{Q}$ satisfying

(1) $\quad\quad\quad\quad\quad\quad$ **L** $\cap$ **R** = $\emptyset$
(2) $\quad\quad\quad\quad\quad\quad \forall p[p \in$ **L** $\leftrightarrow \exists q \in$ **L**$.p < q]$
(3) $\quad\quad\quad\quad\quad\quad \forall p[p \in$ **R** $\leftrightarrow \exists q \in$ **R**$.q < p]$
(4) $\quad\quad\quad\quad\quad\quad \forall p \forall q[p < q \to p \in$ **L** $\vee q \in$ **R**$]$

We write $\mathbb{R}_d$ for the set of Dedekind real numbers.

---

[13] See, e.g. Mac Lane and Birkhoff [1967], Chapter II, section 4.
[14] Mac Lane and Birkhoff [1967], Chapter V section 2.

We also define a *weak real number* to be a pair is a pair <**L, R**> of inhabited subsets **L, R** ⊆ ℚ satisfying (1), (2), (3) and the pair of conditions (both weaker than (4)):

(4*) $\quad\quad\quad\quad \forall p \forall q [p < q \wedge p \notin L \to q \in R]$
$\quad\quad\quad\quad\quad \forall p \forall q [p < q \wedge q \notin R \to p \in L]$

We write $\mathbb{R}_w$ for the set of weak real numbers. Clearly $\mathbb{R}_d \subseteq \mathbb{R}_w$. We shall use letters r, s as vas variables ranging over $\mathbb{R}_w$, and write $r = <\mathbf{L}_r, \mathbf{R}_r>$.

Note that, for any weak real number r, $\mathbf{R}_r$ is *recoverable* from $\mathbf{L}_r$ in that

$$\mathbf{R}_r = \{p : \exists q < p (q \notin \mathbf{L}_r)\}.$$

The *ordering* ≤ on $\mathbb{R}_w$ (and its restriction to $\mathbb{R}_d$) is defined by [15]

$$r \leq s \leftrightarrow \mathbf{L}_r \subseteq \mathbf{L}_s.$$

The *strong ordering* < on $\mathbb{R}_w$ (and its restriction to $\mathbb{R}_d$) is defined by

$$r < s \leftrightarrow \exists p (p \in \mathbf{R}_r \wedge p \in \mathbf{L}_s).;$$

Clearly ≤ is a (partial) ordering on $\mathbb{R}_w$ (hence also on $\mathbb{R}_d$), and it is straightforward to show that < is irreflexive and transitive.

While classically it can be shown that $r \leq s \leftrightarrow r < s \vee r = s$, this does not hold in **IZ**. What can be proved is

Lemma 1. (i) *In* $\mathbb{R}_w$, $r \leq s \leftrightarrow \neg(s < r)$.
(ii) *In* $\mathbb{R}_d$, $r < s \to \forall e (r < e \vee e < s)$. [16]

---

[15] The asymmetry in this definition is only apparent since from the recoverability of **R** from **L**, noted above, follows that $\mathbf{L}_r \subseteq \mathbf{L}_s \leftrightarrow \mathbf{R}_s \subseteq \mathbf{R}_s$.
[16] This condition does not necessarily hold in $\mathbb{R}_w$.

Proof. (i) From $r \leq s$ and $s < r$ we deduce $\exists p(p \in \mathbf{R}_s \wedge p \in \mathbf{L}_s)$, contradicting (1), so $r \leq s \to \neg(s < r)$. Conversely, $\neg(s < r)$ is equivalent to $\forall p \neg (p \in \mathbf{R}_s \wedge p \in \mathbf{L}_r)$. So from $q \in \mathbf{L}_r$ it follows from (2) that $\exists p(p > q \wedge p \in \mathbf{L}_r)$, whence $\exists p(p > q \wedge p \notin \mathbf{R}_s)$. Thus by (4*) $q \in \mathbf{L}_s$. So $\mathbf{L}_s \subseteq \mathbf{L}_s$ and $r \leq s$.

(ii). If $r, s \in \mathbb{R}_d$ and $r < s$, we get rational $p < q$ for which $p \in \mathbf{R}_r \wedge q \in \mathbf{L}_s$. If now $e \in \mathbb{R}_d$, then $p \in \mathbf{L}_e \vee q \in \mathbf{R}_e$. It follows that

$$(p \in \mathbf{R}_r \wedge p \in \mathbf{L}_e) \vee (q \in \mathbf{L}_s \wedge q \in \mathbf{R}_e).$$

The first disjunct implies $r < e$ and the second $e < s$. ∎

In the classical case one now proceeds to show that $\mathbb{R}_d$ is *conditionally order-complete*, i.e. every inhabited subset with an upper bound has a least upper bound. The argument for the conditional order-completeness of $\mathbb{R}_d$ requires an application of **LEM** which is not available in **IZ**. On the other hand, we shall prove in **IZ** that $\mathbb{R}_w$ *is conditionally order-complete*. Moreover, we shall show in **IZ** that, if De Morgan's Law **DML** $\neg(\varphi \wedge \psi) \to (\neg\varphi \vee \neg\psi)$ holds, then $\mathbb{R}_d$ is conditionally order-complete, and conversely.

Proposition 1. *In* **IZ**, $\mathbb{R}_w$ *is conditionally order-complete.*

Proof. Let X be a bounded inhabited subset of $\mathbb{R}_w$. We want to construct a least upper bound $r_X = \langle \mathbf{L}, \mathbf{R} \rangle$ for X. Define **R, L** by

$$\mathbf{R} = \{q : \exists p < q \forall r \in X (p \in \mathbf{R}_r)\}$$
$$\mathbf{L} = \{p : \exists q (p < q \wedge q \notin \mathbf{R})\}.$$

It is easy to verify that r_X satisfies conditions (1), (2), (3) and the second condition in (4*) above. To verify the first condition in (4*), suppose that $p < q \wedge p \notin \mathbf{L}$. Define

$$p' = \frac{2p+q}{3}, \quad q' = \frac{p+2q}{3}$$

Then $p < p' < q' < q$. From the definition of **L** it follows that $\neg\neg p' \in \mathbf{R}$. But since $\mathbf{R} \subseteq \mathbf{R}_r$ for all $r \in X$, it must be the case that $\mathbf{R} \cap \mathbf{L}_r = \emptyset$ for all $r \in X$, whence $p' \notin \mathbf{L}_r$ for all $r \in X$. It follows that $\forall r \in X(q' \in \mathbf{R}_r)$, so that $q \in \mathbf{R}$.

Since $\mathbf{L}_r \subseteq \mathbf{L}$ for all $r \in X$, r_X is an upper bound for X. But if s is any upper bound for X, then $\mathbf{R}_s \subseteq \mathbf{R}_r$ for all $r \in X$, and it follows easily from this that $\mathbf{R}_s \subseteq \mathbf{R}$. Hence $s \leq r_X$ and we are done. ∎

Proposition 2. *The following are equivalent in* **IZ**:

(i)   DML (or, equivalently, WLEM)
(ii) $\mathbb{R}_d = \mathbb{R}_w$
(iii) $\mathbb{R}_d$ *is conditionally order-complete.*

Proof. (i) → (ii). Assume (i) and let $r \in \mathbb{R}_w$. To show that $r \in \mathbb{R}_d$, suppose given rationals p, q with $p < q$ and let $e = \frac{1}{2}(p + q)$. From the disjointness of $\mathbf{L}_r$ and $\mathbf{R}_r$ together with **DML**, it follows that $e \notin \mathbf{R}_r \lor e \notin \mathbf{L}_r$. But since $p < e$ the first disjunct implies $p \in \mathbf{L}_r$ and the second similarly implies $q \in \mathbf{R}_r$. It follows that $r \in \mathbb{R}_d$, whence (ii).

(ii) → (iii) is an immediate consequence of Proposition 1.

(iii) → (i). Assume (iii) and let φ be any formula. Then the set

$$S = \{q \in \mathbb{Q} : q = 0 \lor [q = 1 \land \varphi]\}$$

is inhabited and bounded in $\mathbb{R}_d$. Let r be the supremum of S. Then we have $\varphi \to r = 1$ and $\neg \varphi \to r = 0$, so that $r < 1 \to \neg \varphi$ and $0 < r \to \neg \neg \varphi$. By (ii) of Lemma 1, $0 < r \lor r < 1$, and we deduce that $\neg\neg\varphi \lor \neg\varphi$. WLEM and hence DML follows. ∎

We next show how the operations of addition and multiplication may be defined on $\mathbb{R}_d$ so as to turn it into a communicative ring[17]. Here we shall only provide a sketch.

Addition on $\mathbb{R}_d$ is defined by

$$<L_1,R_1> + <L_2,R_2>$$
$$= \{p_1 + p_2 : p_1 \in L_1 \wedge p_2 \in L_2\}, \{p_1 + p_2 : p_1 \in R_1 \wedge p_2 \in R_2\}.$$

The definition of multiplication is more involved, since without **LEM** we cannot, as we can classically, divide into cases according to the signs of the multipliers. In fact we first define the product of a real number and a rational, which can be divided into cases since the usual ordering on the rationals satisfies the trichotomy law. Thus we define:

$$p \cdot r = \begin{cases} \langle \{p \cdot q : q \in L_r\}, \{p \cdot q : q \in R_r\} \rangle & \text{if } p > 0 \\ \langle \{p \cdot q : q \in R_r\}, \{p \cdot q : q \in L_r\} \rangle & \text{if } p < 0 \\ 0 & \text{if } p = 0. \end{cases}$$

To define the product of $r_1, r_2 \in \mathbb{R}_d$ we use the idea that if $p_1 - r_1$ and $p_2 - r_2$ have the same sign then $(r_1 - p_1) \cdot (r_2 - p_2)$ should be positive. Thus we define $L_{r_1 \cdot r_2}$ to be

$$\{p \in \mathbb{Q} : \exists p_1 p_2 [((p_1 \in L_{r_1} \wedge p_2 \in L_{r_2}) \vee (p_1 \in R_{r_1} \wedge p_2 \in R_{r_2})) \\ \wedge (p + p_1 \cdot p_2 \in L_{p_2 \cdot r_1 + p_1 \cdot r_2})]\}$$

and $L_{r_1 \cdot r_2}$ to be

$$\{p \in \mathbb{Q} : \exists p_1 p_2 [((p_1 \in R_{r_1} \wedge p_2 \in L_{r_2}) \vee (p_1 \in L_{r_1} \wedge p_2 \in R_{r_2})) \\ \wedge (p + p_1 \cdot p_2 \in R_{p_2 \cdot r_1 + p_1 \cdot r_2})]\}$$

---

[17] In fact the procedure introduced here works equally well for $\mathbb{R}_w$.

A tedious verification shows that these definitions of addition and multiplication do indeed turn $\mathbb{R}_d$ into a commutative ring.

Finally, we discuss the *Cauchy reals*. These are obtained as equivalence classes of Cauchy sequences of rationals and, in classical set theory, the resulting field can be shown to be isomorphic to $\mathbb{R}_d$. As we shall see in Chapter IV this is not necessarily the case in **IZ**.

A *Cauchy sequence* is a function $f \in \mathbb{Q}^\mathbb{N}$ such that

$$\forall m \in \mathbb{N} \forall n \in \mathbb{N}[0 < m \leq n \to | f(m) - f(n) | < 1/m].$$

Let C be the set of Cauchy sequences. We define the relation $E \subseteq C \times C$ of "converging to the same limit" by

$$E = \{< f, g > \in C \times C : \forall n > 0[\, | f(n) - g(n) | < 3/n\}.$$

It can then be shown that E is an equivalence relation on C. To establish the transitivity of E, from $< f, g > \in E, < g, h > \in E$ and $n > 0$, one derives

$$| f(n) - f(6n) | < 1/n, \quad | f(6n) - g(6n) | < 1/2n, \quad | g(6n) - h(6n) | < 1/2n,$$
$$| h(6n) - h(n) | < 1/n,$$

from which the desired inequality follows.

The set $\mathbb{R}_c$ of *Cauchy real numbers* is defined to be the quotient of C by the equivalence relation E, i.e. the set of E-equivalence classes of C.

The operations of addition and multiplication on $\mathbb{R}_c$ are introduced by first defining them on C:

$$(f + g)(n) = f(2n) + g(2n),$$

and

$$(f \cdot g)(n) = f(kn) + g(kn),$$

where $k = |f(1)| + |g(1)| + 3$. It can then be checked that these operations are compatible with E and so induce operations on $\mathbb{R}_c$ which give the latter the structure of a commutative ring.

We define a map $i: C \to \mathbb{R}_d$ by

$$i(f) = \langle \{p : \exists n[p < f(n) - 1/n]\}, \{p : \exists n[p > f(n) + 1/n]\} \rangle.$$

(To verify that $i(f)$ is indeed a Dedekind real, we note that for given $p < q$ in $\mathbb{Q}$ we can choose n with $3/n < q - p$, and then we must have either $p < f(n) - 1/n$ or $q > f(n) + 1/n$.) Moreover, it is not hard to show that

$$\forall f, g \in C[i(f) = i(g) \leftrightarrow < f, g > \in E],$$

so that i induces a monic map $j: \mathbb{R}_c \to \mathbb{R}_d$. This map is a ring homomorphism.

Now j is not necessarily an isomorphism[18]. But it is one in the presence of the *countable axiom of choice* $\mathbf{AC}(\mathbb{N})$[19]. For from $\mathbf{AC}(\mathbb{N})$ it can be deduced that every Dedekind real is the limit of a Cauchy sequence of rationals, since for each Dedekind real r and each $n > 0$ we can find a rational p with $|r - p| < 1/n$. This follows from the fact that, since $\mathbf{L}_r$ and $\mathbf{R}_r$ are inhabited, we can find rationals p, q for which $p < r < q$; then the interval $[p, q]$ can be divided into finitely many subintervals of length $< 1/n$.

---

[18] See Ch. IV.
[19] See Ch. IV.

Chapter IV

Intuitionistic Zermelo-Fraenkel Set Theory and Frame-Valued Models

INTUITIONISTIC ZERMELO-FRAENKEL SET THEORY IZF

Intuitionistic Zermelo-Fraenkel set theory **IZF** is obtained by adding to **IZ** the axioms of *collecction* and $\in$- *induction*

Collection $\forall u[\forall x \in u \exists y \varphi(x,y) \to \exists v \forall x \in u \exists y \in v \varphi(x,y)]$.

$\in$-Induction[20] $\forall x[\forall y \in x \varphi(y) \to \varphi(x)] \to \forall x \varphi(x)$.

It is to be expected that the many classically equivalent definitions of *well-ordering* and *ordinal* become distinct within **IZF**. The definitions we give here work reasonably well.

Definition. A set x is *transitive* if $\forall y \in x\ y \subseteq x$; an *ordinal* is a transitive set of transitive sets. The class of ordinals is denoted by ORD and we use letters $\alpha, \beta, \gamma,..$ as variables ranging over it. A transitive subset of an ordinal is called a *subordinal*. An ordinal α is *simple* if $\forall \beta \in \alpha \gamma \in \alpha (\beta \in \gamma \vee \beta = \gamma \vee \gamma \in \beta)$.

Thus, for example, the ordinals 0, 1, 2, 3, ... as well as the first infinite ordinal to be defined below, are all simple. Every subordinal (hence every element) of a simple ordinal is simple. But, in contrast with classical set theory,

---

[20] In classical set theory the $\in$-induction scheme is equivalent to the *axiom of regularity*, which asserts that each nonempty set u has a member x which is $\in$-*minimal*, that is, for which $x \cap u = \emptyset$. It is easy to see that this implies **LEM**: an $\in$-minimal element of the set $\{0 \mid \varphi\} \cup \{1\}$ is either 0 or 1; if it is 0, φ must hold, and if it is 1, φ must fail; thus if foundation held we would get $\varphi \vee \neg \varphi$.

intuitionistically not every ordinal can be simple, because the simplicity of the ordinal $\{0, \{0 \mid \varphi\}\}$ implies $\varphi \vee \neg\varphi$.

We next state the central properties of ORD.

Definition. The *successor* α^+ of an ordinal α is $\alpha \cup \{\alpha\}$; the *supremum* of a set A of ordinals is $\bigcup A$. The usual *order relations* are introduced on ORD:

$$\alpha < \beta \leftrightarrow \alpha \in \beta \qquad \alpha \leq \beta \leftrightarrow \alpha \subseteq \beta.$$

It is now easily shown that successors and suprema of ordinals are again ordinals and that

$$\alpha < \beta \leftrightarrow \alpha^+ \leq \beta \qquad \bigcup A \leq \beta \leftrightarrow \forall \alpha \in A.\ \alpha < \beta \leq \gamma \to \alpha < \gamma.$$

But straightforward arguments show that any of the following assertions (for arbitrary ordinals α, β, γ) implies **LEM**:

(i) $\alpha < \beta \vee \alpha = \beta \vee \beta < \alpha$,
(ii) $\alpha \leq \beta \vee \beta \leq \alpha$,
(iii) $\alpha \leq \beta \to \alpha < \beta \vee \alpha = \beta$,
(iv) $\alpha < \beta \to \alpha^+ < \beta \vee \alpha^+ = \beta$,
(v) $\alpha \leq \beta < \gamma \to \alpha < \gamma$.

Notice that as a special case of $\in$-induction we have the *Principle of Induction on Ordinals*, namely,

$$\forall \alpha [\forall \beta < \alpha \varphi(\beta) \to \varphi(\alpha)] \to \forall \alpha \varphi(\alpha).$$

Definition. An ordinal α is a *successor* if $\exists \beta\ \alpha = \beta^+$, a *weak limit* if $\forall \beta \in \alpha\ \exists \gamma \in \alpha\ \beta \in \gamma$, and a *strong limit* if $\forall \beta \in \alpha\ \beta^+ \in \alpha$.

Note that both the following assertions imply **LEM**: **(i)** *every ordinal is zero, a successor, or a weak limit*, **(ii)** *all weak limits are strong limits*. For **(i)** this follows from the observation that, for any formula φ, if the specified disjunction applies to the ordinal $\{0 \mid \varphi\}$, then $\varphi \vee \neg\varphi$. As for assertion **(ii)**, define

$$1\varphi = \{0 \mid \varphi\}, \; 2\varphi = \{0, 1\varphi\}, \; \beta = \{0, 1\varphi, 2\varphi, 2\varphi+, 2\varphi++, \ldots\}.$$

Then β is a weak limit, but a strong one only if $\varphi \vee \neg\varphi$.

As in classical set theory, in **IZF** a connection can be established between the class of ordinals and certain natural notions of well-founded or well-ordered structure. Thus a *well-founded* relation on a class A is a binary relation $\prec$ on A which is *inductive*, that is, for each $a \in A$, the class $\{x \in A : x \prec a\}$ is a set and, for every class X such that $X \subseteq A$ we have

$$\forall x \in A [\forall y \prec x (y \in X) \to x \in X] \to X = A.$$

A well-founded relation has no infinite descending sequences and so is irreflexive. Note that the $\in$- induction axiom asserts that $\in$ is a well-founded relation on V. Also, the relation $<$ on ORD is well-founded.

The usual proof in classical **ZF** to justify *definitions by recursion* on a well-founded relation does not use **LEM**, and so is valid in **IZF**. Thus, given a well-founded relation $\prec$ on a class A and a function $F : A \times V \to V$, it is provable in **IZF** that there exists a unique function $G: A \to V$ such that, for any $u \in A$ we have

$$G(u) = F(\langle u, G \mid \{x \in A : x \prec u\}\rangle.)$$

Recursion on the well- founded relation $\in$ will be called $\in$-*recursion,* and recursion on the well-founded relation $<$ on ORD *ordinal recursion.*

We make the following

Definition. If $\prec$ is a well-founded relation on a class A, the associated *rank function* $rk_\prec : A \to \text{ORD}$ is the (unique) function such that for each $x \in A$,

$$rk_\prec(x) = \bigcup_{y \prec x} rk_\prec(y)^+.$$

When $\prec$ is $\in$ restricted to an ordinal, it is easy to see that the associated rank function is the identity.

To obtain a characterization of the *order-types* represented by ordinals we make the following

Definition. A binary relation $\prec$ on a set A is *transitive* if

$$\forall x \in A \forall y \in A \forall z \in A[x \prec y \wedge y \prec z \rightarrow x \prec z]$$

and *extensional* if

$$\forall x \in A \forall y \in A[\forall z(z \prec x \leftrightarrow z \prec y) \rightarrow x = y.$$

A *well-ordering* is a transitive, extensional well-founded relation.

It is easily shown that the well-orderings are exactly those relations isomorphic to $\in$ restricted to some ordinal. For it follows immediately from the axioms of $\in$-induction and extensionality that the $\in$-relation well-orders every ordinal. And conversely, it is easy to prove by induction that the associated rank function on any well-ordering is an isomorphism.

Definition. The *rank function* $\rho: V \rightarrow \text{ORD}$ is defined by $\in$-recursion through the equation

$$\rho(x) = \bigcup_{y \in x} \rho(y)^+.$$

The *cumulative hierarchy* V_α for $\alpha \in \text{ORD}$ is defined by ordinal recursion through the equation

$$V_\alpha = \bigcup_{\beta < \alpha} V_\beta.$$

Proposition 1.

(i) $\forall x\ \rho(x) \in \text{ORD}$

(ii) $x \in y \rightarrow \rho(x) < \rho(y)$.

(iii) $\alpha \leq \beta \to V_\alpha \subseteq V_\beta$.
(iv) $x \subseteq y \subseteq V_\alpha \to x \in V_\alpha$.
(v) $\forall \alpha \; \rho(\alpha) = \alpha$.
(vi) $\forall x \; x \in V_{\rho(x)^+}$.

Proof. The proofs are straightforward inductions. To illustrate, we prove **(vi)**. Suppose $\forall y \in x \; (y \in V_{\rho(y)^+})$. Then $x \subseteq \bigcup_{y < x} V_{\rho(y)^+} \subseteq V_{\rho(x)}$. Hence $x \in \mathbf{P}V_{\rho(x)} = V_{\rho(x)^+}$. ∎

Notice that it follows from **(vi)** that $\forall x \exists \alpha (x \in V_\alpha)$.

FRAME-VALUED MODELS OF IZF DEVELOPED IN IZF

Throughout this section, we argue in **IZF**.

Let H be a frame[21] with top element $\top$ and bottom element $\bot$. An *H-valued structure* is a triple $\mathfrak{S} = \langle S, [\cdot = \cdot], [\cdot \in \cdot] \rangle$ where S is a class and $[\cdot = \cdot]$, $[\cdot \in \cdot]$ are maps $S \times S \to H$ satisfying the conditions

$$[u = u] = \top$$
$$[u = v] = [v = u]$$
$$[u = v] \wedge [v = w] \leq [u = w]$$
$$[u = v] \wedge [u \in w] \leq [v \in w]$$
$$[v = w] \wedge [u \in v] \leq [u \in w]$$

for $u, v, w \in S$.

Let $\mathcal{L}(S)$ be the language obtained from $\mathcal{L}$ by adding a name for each element of S. For convenience we identify each element of S with its name in $\mathcal{L}(S)$ and use the same symbol for both. The maps $[\cdot = \cdot]$, $[\cdot \in \cdot]$ can be extended to a map $[\sigma]$ defined on the class of all $\mathcal{L}(S)$-sentences recursively by:

$$[\sigma \wedge \tau] = [\sigma] \wedge [\tau] \quad [\sigma \vee \tau] = [\sigma] \vee [\tau] \quad [\sigma \Rightarrow \tau] = [\sigma] \Rightarrow [\tau] \quad [\neg \sigma] = [\sigma]^*$$

---

[21] For the definition of frame, see the Appendix.

$$[\exists x \varphi(x)] = \bigvee_{u \in S}[\varphi(u)] \quad [\forall x \varphi(x)] = \bigwedge_{u \in S}[\varphi(u)]$$

For each sentence σ, $[\sigma]$ is called the *truth value* of σ in 𝕰; σ is *true*, or *holds* in 𝕰, written 𝕰 ⊨ σ, if $[\sigma] = \top$ and it is *false* in 𝕰 if $[\sigma] = \bot$. We also write 𝕰 ⊭ σ for $[\sigma] \neq \top$: this means that while σ is not necessarily false in 𝕰, it nevertheless fails to be true in 𝕰. In this event we say that σ is *not affirmed* in 𝕰. It is not hard to show that all the axioms of first-order intuitionistic logic with equality hold in 𝕰, and all its rules of inference are, in the evident sense, valid in 𝕰. 𝕰 is a *(frame-valued) model* of a set T of $\mathcal{L}(S)$-sentences if each member of T is true in 𝕰. If 𝕰 is a model of T, and σ is an intuitionistic consequence of T, then 𝕰 ⊨ σ.

Given a frame H, we set about constructing, within **IZF**, an H-valued structure 𝕍$^{(H)}$ called the *universe of H-sets* or the *H-extension of the universe of sets*[22], which can be proved, in **IZF**, to be itself a frame-valued model of **IZF**. It follows that any sentence σ which is *true* in some 𝕍$^{(H)}$ must be *consistent* with **IZF**.

The *class* $V^{(H)}$ *of H- sets* is defined as follows. First, we define by ordinal recursion the sets $V_\alpha^{(H)}$ for each ordinal α:

$$V_\alpha^{(H)} = \{x : \text{Fun}(x) \wedge \text{ran}(x) \subseteq H \wedge \exists \xi < \alpha [\text{dom}(x) \subseteq V_\xi^{(H)}]\}.$$

Then we define

$$V^{(H)} = \{x : \exists \alpha [x \in V_\alpha^{(H)}]\}.$$

It is easily seen that an H- set is precisely an H-valued function whose domain is a set of H- sets. We write $\mathcal{L}^{(H)}$ for the language $\mathcal{L}(V^{(H)})$.

The basic principle for establishing facts about H- sets is the

Induction Principle for $V^{(H)}$ .. *For any formula* φ(x), *if*

---

[22] When H is the frame $\mathcal{O}(X)$ of open sets in a topological space X,, 𝕍$^{(H)}$ is called a *spatial extension*.

$$\forall x \in V^{(H)}[\forall y \in \mathrm{dom}(x)\, \varphi(y) \to \varphi(x)],$$

then

$$\forall x \in V^{(H)}\, \varphi(x).$$

This is easily proved by induction on rank.

We now proceed to turn $V^{(H)}$ into an H-valued structure. This is done by defining $[\![u = v]\!]^{(H)}$ and $[\![u \in v]\!]^{(H)}$ by $\in$-recursion as follows[23]:

$$[\![u \in v]\!]^{(H)} = \bigvee_{y \in \mathrm{dom}(v)} v(y) \wedge [\![u = y]\!]^{(H)}$$

$$[\![u = v]\!]^{(H)} = \bigwedge_{x \in \mathrm{dom}(u)} [u(x) \Rightarrow [\![x \in v]\!]^{(H)}] \wedge \bigwedge_{y \in \mathrm{dom}(v)} [v(y) \Rightarrow [\![y \in u]\!]^{(H)}].$$

It can now be shown by $\in$-induction that $\mathfrak{B}^{(H)} = \langle V^{(H)}, [\![\cdot = \cdot]\!]^{(H)}, [\![\cdot \in \cdot]\!]^{(H)} \rangle$ is an H-valued structure. This structure is called the *universe of H-sets*: a structure of the form $\mathfrak{B}^{(H)}$ is called a *frame-valued universe*. We assume that $[\![\cdot]\!]^{(H)}$ has been extended to the class of all $\mathcal{L}^{(H)}$ - sentences as above: we shall usually omit the superscript $^{(H)}$.

In particular we have

$$[\![\exists x \varphi(x)]\!] = \bigvee_{u \in V^{(H)}} [\![\varphi(u)]\!] \qquad [\![\forall x \varphi(x)]\!] = \bigwedge_{u \in V^{(H)}} [\![\varphi(u)]\!].$$

Note that we can always find an ordinal α for which

(*) $$[\![\exists x \varphi(x)]\!] = \bigvee_{u \in V_\alpha^{(H)}} [\![\varphi(u)]\!] \qquad [\![\forall x \varphi(x)]\!] = \bigwedge_{u \in V_\alpha^{(H)}} [\![\varphi(u)]\!]$$

For let $A = \{[\![\varphi(u)]\!] : u \in V^{(H)}\}$. Then $A \subseteq H$, and since H is a set, by Separation so is A. We then have

---

[23] For the details in the Boolean-valued case, which are the same as in the frame-valued case, see Bell [2011].

$$\forall x \in A \exists \beta \exists u \in V_\beta^{(H)} (x = \llbracket \varphi(u) \rrbracket).$$

So, using Collection there, is a set $U \subseteq \mathrm{ORD}$ such that

$$\forall x \in A \exists \beta \in U \exists u \in V_\beta^{(H)} (x = \llbracket \varphi(u) \rrbracket).$$

If we let $\alpha = \bigcup U$, then

$$\forall x \in A \exists \beta \in U \exists u \in V_\alpha^{(H)} (x = \llbracket \varphi(u) \rrbracket).$$

so that $A = \{ \llbracket \varphi(u) \rrbracket : u \in V_\alpha^{(H)} \}$ and (*) follows.

An argument of this general sort will be called a *Collection argument*; we shall tacitly employ a number of such arguments in the sequel.

Of use in calculating truth values in $\mathbf{V}^{(H)}$ are the rules:

$$\llbracket \exists x \in u \varphi(x) \rrbracket = \bigvee_{x \in \mathrm{dom}(u)} \llbracket \varphi(x) \rrbracket \qquad \llbracket \forall x \in u \varphi(x) \rrbracket = \bigwedge_{x \in \mathrm{dom}(u)} \llbracket \varphi(x) \rrbracket$$

$$u(x) \le \llbracket x \in u \rrbracket \quad \text{for } x \in \mathrm{dom}(u)$$

Note that, given an H-set u, $\mathbf{V}^{(H)} \models \exists x (x \in u)$ does not necessarily imply that there is an H-set v for which $\mathbf{V}^{(H)} \models v \in u$. An H-set u satisfying this latter condition is called *inhabited*, and an H-set v satisfying $\mathbf{V}^{(H)} \models v \in u$ is called a *definite element* of u.

There is a natural map $\hat{\cdot} : V \to V^{(H)}$ defined by $\in$-recursion as follows:

$$\hat{x} = \{ <\hat{y}, \top> : y \in x \}.$$

Thus $\mathrm{dom}(\hat{x}) = \{\hat{y} : y \in x\}$ and $\hat{x}(\hat{y}) = \top$ for $y \in x$.

It is then easily shown that, for $x \in V$, $u \in V^{(H)}$,

$$[u \in \hat{x}] = \bigvee_{y \in x}[u = \hat{y}].$$

It follows that

$$a \in b \to [\hat{a} \in \hat{b}] = \top$$

And

$$[\exists x \in \hat{a}\ \varphi(x)] = \bigvee_{x \in a}[\varphi(\hat{x})] \qquad [\forall x \in \hat{a}\ \varphi(x)] = \bigwedge_{x \in a}[\varphi(\hat{x})]$$

It follows immediately from these that the H-valued set $\hat{\varnothing}$ represents the natural number 0 (i.e. $\varnothing$) in $\mathcal{B}^{(H)}$. Moreover, the H-valued set $\hat{\mathbb{N}}$ represents the set of *natural numbers* in $\mathcal{B}^{(H)}$. For let $Ind(u)$ be the formula $0 \in u \wedge \forall x \in u(x^+ \in u)$. It is then easily checked that $\mathcal{B}^{(H)} \models Ind(\hat{\mathbb{N}})$. Also, we have, for each $n \in \mathbb{N}$

(*) $$[Ind(u)] \leq [\hat{n} \in u].$$

This is proved by induction on n. It is clearly satisfied by 0, and obviously

$$[Ind(u) \wedge \hat{n} \in u] \leq [\widehat{n^+} \in u].$$

Therefore

$$[Ind(u)] \leq [\hat{n} \in u] \to [Ind(u)] \leq [\widehat{n^+} \in u].$$

and (*) follows by induction. Hence

$$[Ind(u)] \leq \bigwedge_{n \in \mathbb{N}}[\hat{n} \in u] = [\hat{\mathbb{N}} \subseteq u].$$

So $\hat{\mathbb{N}}$ is the least inductive set in $\mathcal{B}^{(H)}$.

Note that it follows in particular that *the Axiom of Infinity holds in* $\mathbf{V}^{(H)}$.

A useful fact is the

Unique Existence Principle for $\mathbf{V}^{(H)}$. *If* $\mathbf{V}^{(H)} \models \exists! x \varphi(x)$, *then* $\mathbf{V}^{(H)} \models \varphi(u)$ *for some* $v \in V^{(H)}$.

Proof. This is proved by translating to $\mathbf{V}^{(H)}$ the proof in IZF that, if $\exists! x \varphi(x)$, then the set u defined by $u = \{y : \exists x [\varphi(x) \wedge y \in x]\}$ satisfies $\varphi(u)$. Thus, assuming $\mathbf{V}^{(H)} \models \exists! x \varphi(x)$, we have $\top = [\![\exists x \varphi(x)]\!] = \bigvee_{x \in V^{(H)}} [\![\varphi(x)]\!]$. Using a Collection argument we obtain an ordinal α for which $\top = \bigvee_{x \in V_\alpha^{(H)}} [\![\varphi(x)]\!]$. If we now define $u \in V^{(H)}$ by $\mathrm{dom}(u) = V_\alpha^{(H)}\}$, $u(y) = [\![\exists x [\varphi(x) \wedge y \in x]\!]$, then $[\![\varphi(u)]\!] = \top$. ∎

As we have observed, it is not in general true that if $\mathbf{V}^{(H)} \models \exists x \varphi(x)$, then $\mathbf{V}^{(H)} \models \varphi(u)$ for some $u \in V^{(H)}$. As we show below, certain conditions on H will ensure that this holds.

Given subsets $\{a_i : i \in I\} \subseteq H, \{u_i : i \in I\} \subseteq V^{(H)}$, we define the *mixture* $\sum_{i \in I} a_i \cdot u_i$ to be the H-set u defined by $\mathrm{dom}(u) = \bigcup_{i \in I} \mathrm{dom}(u_i)$ and, for $x \in \mathrm{dom}(u)$, $u(x) = \bigvee_{i \in I} a_i \wedge [\![x \in u_i]\!]$. If $I = \{0, 1\}$, we write $a_0 \cdot u_0 + a_1 \cdot u_1$ for $\sum_{i \in I} a_i \cdot u_i$.

Two elements of a frame are *disjoint* if their meet is $\bot$. A subset of a frame consisting of mutually disjoint elements is called an *antichain*. If an antichain is presented as an indexed set $\{a_i : i \in I\}$, we shall always assume that $a_i \wedge a_{i'} = \bot$ whenever $i \neq i'$.

Now we can prove the

Mixing Lemma. *Let* $\{u_i : i \in I\} \subseteq V^{(H)}$ *and suppose that* $\{a_i : i \in I\}$ *is an antichain in* H. *Then, writing u for* $\sum_{i \in I} a_i \cdot u_i$, *we have* $a_i \leq [\![u = u_i]\!]$ *for all* $i \in I$.

Proof. We have, for given $i \in I$, $[u = u_i] = a \wedge b$, where

$$a = \bigwedge_{x \in \text{dom}(u)} u(x) \Rightarrow [x \in u_i] \qquad b = \bigwedge_{x \in \text{dom}(u_i)} u_i(x) \Rightarrow [x \in u]$$

If $x \in \text{dom}(u)$, then, since $\{a_i : i \in I\}$ is an antichain,

$$a_i \wedge u(x) = \bigvee_{j \in I}(a_i \wedge a_j \wedge [x \in u_j]) = a_i \wedge [x \in u_i] \leq [x \in u_i].$$

Hence $a_i \leq [u(x) \Rightarrow [x \in u_i]]$ for all $x \in \text{dom}(u)$, so that $a_i \leq a$. On the other hand, if $x \in \text{dom}(u_i)$, then

$$a_i \wedge u_i(x) \leq a_i \wedge [x \in u_i] \leq u(x) \leq [x \in u],$$

so that $a_i \leq [u_i(x) \Rightarrow [x \in u]]$, whence $a_i \leq b$. Hence $a_i \leq a \wedge b$ and the result follows. ■

An element a of a frame H is said to be *complemented* if $a \vee a^* = \top$. H is *totally disconnected* if every element of H s the join of a set of complemented. elements. Equivalently, H is totally disconnected if, for any elements a, b, $a \leq b$ iff, for all complemented elements c, $c \leq a$ implies $c \leq b$. Notice that every Boolean algebra is totally disconnected, and, for a topological space X, the frame $\mathcal{O}(X)$ of open sets in X is totally disconnected iff X s totally disconnected in the topological sense of having a base of clopen sets.

We shall need the following fact later on:

Definite Element Lemma *Suppose that H is totally disconnected, let u be an inhabited H-valued set, and write U for the class of definite elements of u. Then, for any formula $\varphi(x)$,*

(i) $[\forall x \in u \varphi(x)] = \bigwedge_{w \in U} [\varphi(w)]$;

(ii) *the following are equivalent:*
   (a) $\mathbf{V}^{(H)} \models \forall x \in u \varphi(x)$
   (b) $\mathbf{V}^{(H)} \models \varphi(w)$ *for all $u \in U$.*

65

Proof. (i). We first prove the following:

(*) *For any H-set v and any complemented $a \in H$ such that $a \leq [v \in u]$ there is a definite element $w \in U$ such that $a \leq [\![v = w]\!]$.*

Suppose that v and a satisfy the hypothesis. Choose a definite element z of u and let $w = a \cdot v + a^* \cdot z$. Then $a \leq [v = w]$ and $a^* \leq [z = w]$ follow from the Mixing Lemma. Moreover, we have

$$[w \in u] \geq [v = w \wedge v \in u] \vee [z = w \wedge z \in u] \geq a \vee a^* = \top,$$

and so w is a definite element of u. This proves (*)

To prove **(i)** it suffices to show that

$$\bigwedge_{w \in U} [\varphi(w)] \leq [\forall x \in u \varphi(x)]$$

i.e.,

$$\bigwedge_{w \in U} [\varphi(w)] \leq \bigwedge_{v \in V^{(H)}} [[v \in u] \Rightarrow [\varphi(v)]].$$

Thus we must show that, for each $V^{(H)}$-set v,

$$\bigwedge_{w \in U} [\varphi(w)] \leq [[v \in u] \Rightarrow [\varphi(v)]]$$

that is,

$$\bigwedge_{w \in U} [\varphi(w)] \wedge [[v \in u]] \leq [\varphi(v)].$$

Since H s totally disconnected, to prove this it is enough to show that, for any complemented element $a \in H$,

(**) $$a \leq \bigwedge_{w \in U} [\varphi(w)] \wedge [v \in u] \rightarrow a \leq [\varphi(v)].$$

So assume the antecedent of (**). By (*), there is $w \in U$ such that $a \leq [v = w]$. Hence $a \leq [v = w] \wedge [\varphi(w)] \leq [\varphi(v)]$. This proves (**), and **(i)** follows.

(ii). Obviously (a) implies (b). Conversely, if (b) holds, then $\bigwedge_{w \in U} [\varphi(w)] = \tau$ and it now follows from **(i)** that $[\forall x \in u \varphi(u)] = \tau$. ∎

A *core* for an H-set u is a set $C \subseteq V^{(H)}$ such that (i) each member of C is a definite element of u; (ii) for any definite element y of u there is $x \in C$ such that $[x = y] = \tau$.

We show that each H-set u has a core. For each $x \in V^{(H)}$ let

$$a_x = \{\langle z, u(z) \wedge [z = x] \rangle : x \in \mathrm{dom}(u)\}.$$

By a Collection argument there is a set $W \subseteq V^{(H)}$ such that, for any $x \in V^{(H)}$, there is $y \in W$ for which $a_x = a_y$. It is easily shown that the set $C = \{x \in w : [x \in u] = \tau\}$ is a core for u.

By abuse of notation, we shall write $\{x \in V^{(H)} : [x \in u] = \tau\}$ to denote a core for a given H-set u

Now in **IZF** it can be shown that $\mathcal{B}^{(H)}$ *is an H-valued model of* **IZF**. It was shown above that the Axiom of Infinity holds in $\mathcal{B}^{(H)}$. We further verify the Axioms of Separation, Collection and $\in$-Induction in $\mathcal{B}^{(H)}$, with brief comments on the verification of (some of) the remaining axioms.

To begin with, we note that, given H-sets u, v, the H-set $\{u, v\}^{(H)} = \{\langle u, \tau \rangle, \langle v, \tau \rangle\}$ is easily shown to validate the Pairing Axiom in $\mathcal{B}^{(H)}$.

In this connection $\{u\}^{(H)} = \{u, u\}^{(H}$ may be identified as the singleton of u in $\mathcal{B}^{(H)}$, and $<u, v>^{(H)} = \{\{u\}^{(H)}, \{u, v\}^{(H)}\}^{(H)}$ as the *ordered pair* of u, v in $\mathcal{B}^{(H)}$.

We recall that the Axiom of Separation is the scheme

$$\forall u \exists v \forall x [x \in v \leftrightarrow x \in u \wedge \varphi(x)].$$

To see that each instance holds in $\mathbf{V}^{(H)}$, let $u \in V^{(H)}$, define $v \in V^{(H)}$ by dom(v) = dom(u) and, for $x \in$ dom(v), $v(x) = u(x) \wedge [\![\varphi(x)]\!]$. Then we have

$$[\![\forall x [x \in v \leftrightarrow x \in u \wedge \varphi(x)]]\!] = [\![\forall x \in v [x \in u \wedge \varphi(x)]]\!] \wedge [\![\forall x \in u [\varphi(x) \to x \in v]]\!].$$

Now

$$[\![\forall x \in v [x \in u \wedge \varphi(x)]]\!] = \bigwedge_{x \in \text{dom}(v)} [\![u(x) \wedge [\![\varphi(x)]\!]] \Rightarrow [x \in u \wedge [\![\varphi(x)]\!]]\!] = \top.$$

Similarly

$$[\![\forall x \in u [\varphi(x) \to x \in v]]\!] = \top$$

and so Separation holds in $\mathbf{V}^{(H)}$.

As for Collection, we recall that this is

$$\forall u [\forall x \in u \exists y \varphi(x, y) \to \exists v \forall x \in u \exists y \in v \varphi(x, y)].$$

To verify this in $\mathbf{V}^{(H)}$, observe that

$$[\![\forall x \in u \exists y \varphi(x, y)]\!] = \bigwedge_{x \in \text{dom}(u)} [\![u(x) \Rightarrow [\![\exists y \varphi(x, y)]\!]]\!]$$

For each $x \in$ dom(u), there is an ordinal β for which $[\![\exists y \varphi(x, y)]\!] = \bigvee_{y \in V_\beta^{(H)}} [\![\varphi(x, y)]\!]$. So by a Collection argument there is an ordinal α such that, for all $x \in$ dom(u), we have $[\![\exists y \varphi(x, y)]\!] \leq \bigvee_{y \in V_\alpha^{(H)}} [\![\varphi(x, y)]\!]$. If we now define $v \in V^{(H)}$ by $v = V_\alpha^{(H)} \times \{\top\}$, then

$$[\forall x \in u \exists y \varphi(x,y)] = \bigwedge_{x \in \text{dom}(u)} [u(x) \Rightarrow [\exists y \varphi(x,y)]] \leq \bigwedge_{x \in \text{dom}(u)} [u(x) \Rightarrow [\exists y \in v \varphi(x,y)]]$$
$$= [\forall x \in u \exists y \in v \varphi(x,y)].$$

The truth of Collection in $\mathbf{V}^{(H)}$ follows.

The $\in$- induction axiom is, we recall, $\forall x [\forall y \in x \varphi(y) \to \varphi(x)] \to \forall x \varphi(x)$. To verify the truth of this in $\mathbf{V}^{(H)}$, first put

$$a = [\forall x [\forall y \in x \varphi(y) \to \varphi(x)]].$$

It now suffices to show that, for any $x \in V^{(H)}$, $a \leq [\varphi(x)]$. To do this we use the induction principle for $V^{(H)}$. Assume for $y \in \text{dom}(x)$ that $a \leq [\varphi(y)]$. Then

$$a \leq \bigwedge_{y \in \text{dom}(x)} [\varphi(y)] \leq \bigwedge_{y \in \text{dom}(x)} [x(y) \Rightarrow [\varphi(y)]] = [\forall y \in x \varphi(y)].$$

But $a \leq [\forall y \in x \varphi(y)] \Rightarrow [\varphi(x)]$, so that

$$a \leq [[\forall y \in x \varphi(y)] \Rightarrow [\varphi(x)]] \wedge [\forall y \in x \varphi(y)] \leq [\varphi(x)],$$

as required.

To establish the truth of the Axiom of Union in $\mathbf{V}^{(H)}$, given $u \in V^{(H)}$, define $v \in V^{(H)}$ by $\text{dom}(v) = \bigcup_{y \in \text{dom}(u)} \text{dom}(y)$ and, for $x \in \text{dom}(v)$, $v(x) = [\exists y \in u. x \in y]$. It is then easily shown that in $\mathbf{V}^{(H)}$, v is the union of u.

For the Power Set Axiom it can be verified that, in $\mathbf{V}^{(H)}$, the power set of a set $u \in V^{(H)}$ is given by the set $v \in V^{(H)}$ defined by $\text{dom}(v) = H^{\text{dom}(u)}$ and, for $x \in \text{dom}(v)$, $v(x) = [x \in u]$. When u is of the form $\hat{a}$, v may be taken to be the function on $H^{\text{dom}(\hat{a})}$ with constant value ⊤. We write $\mathbf{P}^{(H)}(u)$ for v.

Now it is readily shown that **LEM** holds in $\mathbf{V}^{(H)}$ if and only if H is a Boolean algebra. Since, as we have seen, in **IZ** the Axiom of Choice implies **LEM**, it holds in $\mathbf{B}^{(H)}$; in fact, it cannot hold in $\mathbf{B}^{(H)}$ unless H is a Boolean algebra. This is far from being the case for *Zorn's lemma*, however, despite the fact that it is classically equivalent to **AC**. Indeed, we will show that, in **IZF**, *Zorn's Lemma implies its truth in any* $\mathbf{B}^{(H)}$. We shall take Zorn's lemma in the form: *any inhabited partially ordered set in which every chain has a supremum also has a maximal element.*

Thus suppose $X, \leq_X \in V^{(H)}$ satisfy

$$\mathbf{B}^{(H)} \models\, < X, \leq_X > \text{ is an inhabited partially ordered set in which every chain has a supremum.}$$

Let $X' = \{x \in V^{(H)} : [\![x \in X]\!] = \mathsf{T}\}$ be a core for X and define the relation $\leq_{X'}$ on X by $x \leq_{Y'} x' \leftrightarrow [\![x \leq_X x']\!] = \mathsf{T}$. It is then easily verified that $<X', \leq_{X'}>$ is an inhabited partially ordered set in which every chain has a supremum. So, by Zorn's lemma, X' has a maximal element c. We claim that

(1) $\qquad\qquad [\![c \text{ is a maximal element of } X]\!] = \mathsf{T}.$

To prove (1) we take any $a \in V^{(H)}$ and define $Z \in V^{(H)}$ by $\text{dom}(Z) = \text{dom}(X)$ and

$$Z(x) = [\![x = a \wedge x \in X \wedge c \leq_X x]\!] \vee [\![x = c]\!].$$

for $x \in \text{dom}(X)$. It is then readily verified that $\mathbf{V}^{(H)} \models Z$ *is a chain in* X; and so, using the unique existence principle for $V^{(H)}$, there is $v \in X'$ for which

(2) $\qquad\qquad \mathbf{B}^{(H)} \models v \text{ is the supremum of } Z.$

Since $[\![c \in Z]\!] = \mathsf{T}$ it follows that $[\![c \leq_X v]\!] = \mathsf{T}$, whence $c \leq_{X'} v$, so that $v = c$ by the maximality of c. This and (2) now yield $[\![a \in Z \to a \leq_X c]\!] = \mathsf{T}$; and clearly $[\![a \in V \to c \leq_X x]\!] = \mathsf{T}$. Therefore

(3) $\qquad\qquad [\![a \in Z \to a = c]\!] = \mathsf{T}.$

It is easily verified that

(4) $$[\![a \in X \wedge c \leq_X a]\!] \leq [\![a \in Z]\!].$$

(3) and (4) yield $[\![a \in X \wedge c \leq_X a]\!] \leq [\![a = c]\!]$; since this holds for arbitrary $a \in V^{(H)}$, (1) follows.

From the fact that Zorn's lemma holds in every $\mathbf{B}^{(H)}$ but **AC** does not we may infer that, in **IZF**, the former does not imply the latter. In **IZF** Zorn's lemma is thus very weak, indeed so weak as to be entirely compatible with intuitionistic logic. For more on this see Bell [1997].

THE CONSISTENCY OF ZF AND ZFC RELATIVE TO IZF

We noted above that, when H is a (complete) Boolean algebra, LEM holds in $\mathbf{B}^{(H)}$, so that $\mathbf{B}^{(H)}$ is a model of (classical) **ZF**. In **IZF** the simplest complete Boolean algebra is *not* the two element Boolean algebra 2, since as we have noted above it is complete if and only if **WLEM** holds. The simplest complete Boolean algebra in **IZF** is in fact the Booleanization[24] Ω^{bool} of Ω. Accordingly in **IZF** $\mathbf{B}^{(\Omega^{bool})}$ is a model of **ZF**. It follows that, *if* **IZF** *is consistent, so is* **ZF**. Since the consistency of **ZF** implies (as is well-known) the consistency of **ZFC**, we conclude that, *if* **IZF** *is consistent, so is* **ZFC**.

We shall exploit this last fact in the following way. Suppose we want to show that a certain sentence σ of our set-theoretic language is relatively consistent with **IZF**. We produce a certain frame H and show in **ZF(C)** that σ holds in $\mathbf{B}^{(H)}$. The latter is accordingly a model of **IZF** + σ, from which it follows that the consistency of **ZF(C)** implies that of **IZF** + σ. Since the consistency of **IZF** implies that of **ZF(C)**, we conclude that σ is relatively consistent with **IZF**.

This idea will be used in the following sections.

---

[24] For the definition of the Booleanization of a frame see the Appendix.

FRAME-VALUED MODELS OF IZF DEVELOPED IN ZFC

Henceforth we argue in **ZFC**.

The following additional basic facts concerning frame-valued models can be proved in **ZF**[25].

- $a \in b \leftrightarrow \mathbf{V}^{(H)} \vDash \hat{a} \in \hat{b}$ $a = b \leftrightarrow \mathbf{V}^{(H)} \vDash \hat{a} = \hat{b}$ $a \neq b \rightarrow [\hat{a} = \hat{b}]^{(H)} = \bot$
- if $\varphi(x_1, ..., x_n)$ is a restricted formula[26], then

$$\varphi(a_1, ..., a_n) \leftrightarrow \mathbf{V}^{(H)} \vDash \varphi(\widehat{a_1}, ..., \widehat{a_n}).$$

These facts have certain straightforward consequences which we shall employ without explicit mention, for example:

- $\mathbf{V}^{(H)} \vDash \widehat{\langle a, b \rangle} = \langle \hat{a}, \hat{b} \rangle$
- $f: A \rightarrow B \wedge a \in A \rightarrow \mathbf{V}^{(H)} \vDash \hat{f}: \hat{A} \rightarrow \hat{B} \wedge \hat{f}(\hat{a}) = \widehat{f(a)}$
- $\mathbf{V}^{(H)} \vDash \widehat{B^A} \subseteq \hat{B}^{\hat{A}}$

We shall strengthen the concept of a core for an H-valued set in the following way. A *strong core* for an H-valued set u is a set $v \subseteq V^{(H)}$ such that (i) $[x \in u] = \top$ for all $x \in v$; (ii) for any $y \in V^{(H)}$ such that $[y \in u] = \top$, there is a unique $x \in v$ such that $[x = y] = \top$. It is easy to show, assuming **AC**, that any H-valued set u has a strong core. Starting with a core v for u, define the equivalence relation $\sim$ on v by $x \sim y \leftrightarrow [x = y] = \top$, and let w be a set obtained by selecting one member from each $\sim$-equivalence class. Then w is a strong core for u. Clearly a strong core for an H-valued set is unique up to bijection in the sense that there is a bijection between any pair of such strong cores.

---

[25] For proofs (in the Boolean-valued case) we refer the reader to Bell [2011].
[26] A formula is *restricted* if each of its quantifiers occurs in the form $\forall x \in y$ or $\exists x \in y$ or can be proved equivalent in **IZF** to such a formula.

We shall henceforth write $\{x \in V^{(H)} : [x \in u] = \top\}$ to denote a strong core for a given H-valued set u.

A *refinement* of a subset A of H is a subset B of H such that, for any $b \in B$, there is $a \in A$ such that $b \leq a$. If B is an antichain, it is called a *disjoint refinement* of A. H s called *refinable* if every subset of H has a disjoint refinement with the same join.

We say that $\mathbf{B}^{(H)}$ satisfies the **Existence Principle** if for any formula $\varphi(x)$ there is an H-set u for which $[\exists x \varphi(x)] = [\varphi(u)]$.

We can now prove the

Refinable Existence Lemma. *If H is refinable, $\mathbf{B}^{(H)}$ satisfies the Existence Principle.*

Proof. Suppose that H is refinable. Then for any formula $\varphi(x)$, by a Collection argument there is a subset A of $V^{(H)}$ for which

$$[\exists x \varphi(x)] = \bigvee_{x \in V^{(H)}} [\varphi(x)] = \bigvee_{x \in A} [\varphi(x)].$$

Since H is refinable, $\{[\varphi(x)] : x \in A\}$ has a disjoint refinement $\{a_i : i \in I\}$ with the same join, i.e. $[\exists x \varphi(x)]$. Using **AC**, select for each $i \in I$ an element $x_i \in A$ for which $a_i \leq [\varphi(x_i)]$. Now define u to be the mixture $\sum_{x \in I} a_i . x_i$. Then for each x_i we have $a_i \leq [u = x_i] \wedge [\varphi(x_i)] \leq [\varphi(u)]$, so that

$$[\exists x \varphi(x)] = \bigvee_{x \in A} [\varphi(x)] = \bigvee_{i \in I} a_i \leq [\varphi(u)].$$

Since clearly

$$[\varphi(u)] \leq [\exists x \varphi(x)],$$

we are done. ∎

Remark. The converse to the Refinable Existence Lemma also holds. For suppose $\mathfrak{B}^{(H)}$ satisfies the Existence Principle. Given $A \subseteq H$ define the H-set v by $\mathrm{dom}(v) = \{\hat{a} : a \in A\}$ and $v(\hat{a}) = a$ for $a \in A$. Then there is an H-set u for which $[\exists x(x \in v)] = [u \in v]$. In that case

$$\bigvee A == [\exists x(x \in v)] = [u \in v] = \bigvee_{a \in A} a \wedge [u = \hat{a}],$$

so that $\{a \wedge [u = \hat{a}] : a \in A\}$ is a disjoint refinement of A with the same join as the latter.

We shall need the following propositions.

Proposition 1. *The following assertions are equivalent:*

(i) There is $f \in V^{(H)}$ for which $\mathfrak{B}^{(H)} \models f$ is a surjection from a subset of $\hat{A}$ onto $\hat{B}$.

(ii) There is a subset $\{u_{ab} : a \in A, b \in B\} \subseteq H$ such that, for each $a \in A$, $\{u_{ab}: b \in B\}$ is an antichain, and for each $b \in B$, $\bigvee_{a \in A} u_{ab} = \top$.

Proof. (i) → (ii). Assuming (i) let $f \in V^{(H)}$ be such that

$$\mathfrak{B}^{(H)} \models f \text{ is a function} \wedge \mathrm{dom}(f) \subseteq \hat{A} \wedge \mathrm{ran}(f) = \hat{B}.$$

Define $u_{ab} = [f(\hat{a}) = \hat{b}]$. It is then easily verified that the u_{ab} satisfy (ii).

(ii) → (i). Suppose the u_{ab} satisfy the conditions of (ii). Define $f \in V^{(H)}$ by $\mathrm{dom}(f) = \{\widehat{\langle a,b \rangle} : a \in A, b \in B\}$ and $f(\widehat{\langle a,b \rangle}) = u_{ab}$. Then $\mathfrak{B}^{(H)} \models \mathrm{dom}(f) \subseteq \hat{A}$, the first condition of (ii) gives $\mathfrak{B}^{(H)} \models f$ is a function and the second condition $\mathfrak{B}^{(H)} \models \mathrm{ran}(f) = \hat{B}$. ∎

Call a set Y a *subquotient* of a set X if there is a surjection from a subset of X onto Y. Proposition 1 then has the immediate

Corollary. *If H satisfies condition* **(ii)** *of Prop. 1, then*

$$\mathcal{B}(H) \models \widehat{B} \text{ is a subquotient of } \widehat{A}. \blacksquare$$

An element a of a frame H is *connected* if for any disjoint $b, c \in H$, $a = b \vee c$ implies $b = a$ or $c = a$. H is said to be *connected* if its top element ⊤ is connected. Equivalently, H is connected if whenever ⊤ is the join of an antichain A, then $⊤ \in A$. H is said to be *locally connected* if each of its elements is the join of connected elements; equivalently, if, for any elements a, b of H, $a \leq b$ iff, for all connected elements c, $c \leq a$ implies $c \leq b$.

If X is a topological space, connectedness of an open subset U of X corresponds precisely to connectedness of U as an element of the frame $\mathcal{O}(X)$, and connectedness (resp. local connectedness) of X to connectedness (resp. local connectedness) of $\mathcal{O}(X)$.

Proposition 2. *The following are equivalent:*

(i)   *H is connected;*

(ii)   *for any set u, and $v \in V^{(H)}$, if $\mathcal{B}(H) \models v \in \widehat{u}$, then there is $x \in v$ such that $\mathcal{B}(H) \models v = \widehat{x}$.*

Proof (i) → (ii). Assume (i) and suppose that $\mathcal{B}(H) \models v \in \widehat{u}$. Then $⊤ = [v \in \widehat{u}] = \bigvee_{x \in u}[v = \widehat{x}]$. But since $\{[v = \widehat{x}] : x \in u\}$ is an antichain and H is connected it follows that $[v = \widehat{x}] = ⊤$ for some $x \in u$, which gives (ii).

(ii) → (i). Assume (ii) and let a, b be disjoint elements of H such that $a \vee b = ⊤$. Define the element $v \in V^{(H)}$ by $v = a.\widehat{0} + b.\widehat{1}$. Then $\mathcal{B}(H) \models v \in \widehat{2}$, so that, by (ii), either $a = [v = \widehat{0}] = ⊤$ or $b = [v = \widehat{1}] = ⊤$. (i) follows. ∎

Given sets I, J, let us call the frame H $\perp$-(I, J) *distributive* if, for any subset $\{a_{ij} : i \in I, j \in J\}$ of H such that, for each $i \in I$, $\{a_{ij} : j \in J\}$ is an antichain, we have

$$\bigwedge_{i \in I} \bigvee_{j \in J} a_{ij} = \bigvee_{f \in J^I} \bigwedge_{i \in I} a_{if(i)}.$$

Proposition 3. *The following are equivalent:*

(i) $\mathbf{V}^{(H)} \models \widehat{J}^{\hat{I}} = \widehat{J^I}$

(ii) H *is* $\perp$-(I, J) *distributive.*

Proof. (i) $\to$ (ii). Assume (i), and let $\{a_{ij} : i \in I, j \in J\} \subseteq H$ be such that $\{a_{ij} : j \in J\}$ is an antichain for each $i \in I$. Define $h \in V^{(H)}$ by dom$(h) = \{\widehat{\langle i, j \rangle} : i \in I, j \in J\}$ and $h(\widehat{\langle i, j \rangle}) = a_{ij}$. It is then easily verified that

$$\mathbf{V}^{(H)} \models h \text{ is a function} \wedge \text{dom}(h) \subseteq \hat{I} \wedge \text{ran}(h) \subseteq \hat{J}.$$

Thus $[h(\hat{i}) = \hat{j}] = a_{ij}$ and so

$$\bigwedge_{i \in I} \bigvee_{j \in J} a_{ij} = [\text{dom}(h) = \hat{I}] = [h \in \widehat{J}^{\hat{I}}] = [h \in \widehat{J^I}] = \bigvee_{f \in J^I} [h = \hat{f}]$$

$$= \bigvee_{f \in J^I} \bigwedge_{i \in I} [h(\hat{i}) = \widehat{f(i)}]$$

$$= \bigvee_{f \in J^I} \bigwedge_{i \in I} a_{if(i)}.$$

This is (ii).

(ii) $\to$ (i). Assume (ii). To obtain (i) it suffices to show that

(*) $\mathbf{V}^{(H)} \models \widehat{J}^{\hat{I}} \subseteq \widehat{J^I}$.

For $h \in V^{(H)}$ let $h_{ij} = [\langle \hat{i}, \hat{j} \rangle \in h]$ and $a_{ij} = h_{ij} \wedge \bigwedge_{j \neq j'}[h_{ij} \wedge h_{ij'} \Rightarrow \bot]$. Note that $a_{ij} \wedge a_{ij'} = \bot$ for all $i \in I, j \neq j' \in J$. Then we have

$$[\![h \in \hat{J}^{\hat{I}}]\!] \leq [\![h \text{ is a function} \wedge \operatorname{dom}(f) = \hat{I} \wedge \operatorname{ran}(f) \subseteq \hat{J}]\!]$$
$$\leq \bigwedge_{i \in I} \bigwedge_{j,j' \in J} [h_{ij} \wedge h_{ij'} \Rightarrow [\![\hat{j} = \hat{j'}]\!]] \wedge \bigwedge_{i \in I} \bigvee_{j \in J} h_{ij}$$
$$= \bigwedge_{i \in I} [\bigvee_{j \in J} h_{ij} \wedge \bigwedge_{j,j' \in J} [h_{ij} \wedge h_{ij'} \Rightarrow [\![\hat{j} = \hat{j'}]\!]]$$
$$= \bigwedge_{i \in I} \bigvee_{j \in J} a_{ij} = \bigvee_{f \in J^I} \bigwedge_{i \in I} a_{if(i)} \leq \bigvee_{f \in J^I} \bigwedge_{i \in I} h_{if(i)} = \bigvee_{f \in J^I} \bigwedge_{i \in I} [\![h(\hat{i}) = \widehat{f(i)}]\!]$$
$$= \bigvee_{f \in J^I} [\![h = \hat{f}]\!] = [\![h \in \widehat{J^I}]\!].$$

This proves (*). ■

Let us say that H

- is *completely $\bot$-distributive* if it is $\bot$-(I, J) distributive for all I, J,
- is *completely $\bot$-2 – distributive* if it is $\bot$-$(I, 2)$ distributive for all I, J
- *preserves exponentials* if $\mathfrak{B}^{(H)} \models \hat{J}^{\hat{I}} = \widehat{J^I}$ for all I, J.

Then we have

Proposition 4. *The following are equivalent:*

(i)   *H preserves exponentials*
(ii)   *H is completely $\bot$-distributive*
(iii)   *H is completely $\bot$-2 – distributive*
(iv)   *H is locally connected.*

Proof. The equivalence of **(i)** and **(ii)** follows immediately from Prop. 3. So it suffices to prove the equivalence of **(ii)**, **(iii)** and **(iv)**.

(ii) → (iii). Obvious.

(iii) → (iv). To begin with, for elements $b \leq a$ of H, call b *complemented in a* if there is $c \leq a$ such that $b \wedge c = \bot, b \vee c = a$. It is easy to check that, if such c exists, it is unique; denote it by $a - b$. Now assume (iii), and let $a \in H$. Write $\{a_i : i \in I\}$ for the set of elements of H which are complemented in a. For each $i \in I$ let $a_{i0} = a_i$, $a_{i1} = a - a_i$. Then we have, for each $g \in 2^I$,

(*) $$a = \bigwedge_{i \in I}(a_{i0} \vee a_{i1}) = \bigvee_{f \in 2^I} \bigwedge_{i \in I} a_{if(i)} = \bigwedge_{i \in I} a_{ig(i)} \vee \bigvee_{f \in 2^I - \{g\}} \bigwedge_{i \in I} a_{if(i)}.$$

It follows that each $\bigwedge_{i \in I} a_{ig(i)}$ is complemented in a (with complement $\bigvee_{f \in 2^I - \{g\}} \bigwedge_{i \in I} a_{if(i)}$). We claim that $b = \bigwedge_{i \in I} a_{ig(i)}$ is also connected. For suppose that $b = c \vee d$ with $c \wedge d = \bot$ and $c \neq \top$. Then c is complemented in b and it follows easily that c is complemented in a. Thus $c = a_{i_0}$ for some $i_0 \in I$ and so $\bot \neq c = a_{i_0} \leq a_{i_0 g(i_0)}$. Hence $g(i_0) = 0$, so that $b \leq a_{i_0 g(i_0)} = a_{i_0 0} = a_{i_0} = c$. Accordingly $b = c$ and it follows that b is connected. From (*) we conclude that each $a \in H$ is the join of connected elements, so that H is locally connected.

(iv) → (ii). Suppose that H is locally connected. To show that H is completely $\bot$-distributive, it suffices to show that, for any subset $\{a_{ij} : i \in I, j \in J\}$ of H such that, for each $i \in I$, the set $\{a_{ij} : j \in J\}$ is an antichain, and any connected element $c \in H$, we have

(*) $$c \leq \bigwedge_{i \in I} \bigvee_{j \in J} a_{ij} \to c \leq \bigvee_{f \in J^I} \bigwedge_{i \in I} a_{if(i)}.$$

So assume the antecedent of this implication. Then for each $i \in I$, $c \leq \bigvee_{j \in J} a_{ij}$, and, because c is connected, it follows that $c \leq a_{ij}$ for some unique $j \in J$. Define $g: I \to J$ to be the function which assigns this j to each i. Then $c \leq \bigwedge_{i \in I} a_{ig(i)} \leq \bigvee_{f \in J^I} \bigwedge_{i \in I} a_{if(i)}$, and (*) follows. ∎

A FRAME-VALUED MODEL OF IZF IN WHICH $\mathbb{N}^\mathbb{N}$ IS SUBCOUNTABLE

We now set about constructing (in **ZF**) a frame-valued model of **IZF** in which $\mathbb{N}^\mathbb{N}$ is *subcountable*.

Given two sets A and B, let $P = P(A, B)$ be the set of finite partial functions from A to B, partially ordered by $\supseteq$. We shall write p, q for elements of P. Let **C** be the coverage[27] on P defined by

$$S \in C(p) \leftrightarrow \exists b \in B \ \ S = \{q \in P : p \subseteq q \wedge b \in \mathrm{ran}(q)\}.$$

The sieve S on the left-hand side of this equivalence is called the *cover of p determined by b*. Note that every $S \in C(p)$ is nonempty.

Lemma 1. *For $S \in C(p)$, $q_1, q_2 \in S$, there are $q_3, q_4, q_5 \in S$ such that $q_3 \subseteq q_1, q_4 \subseteq q_2$ and $q_3 \cup q_4 \subseteq q_5$.*

Proof. Suppose S is determined by $b \in B$. The without loss of generality we can assume that $b \notin \mathrm{ran}(p)$, for otherwise we can take $q_3 = q_4 = q_5 = p$. Let $a_1, a_2 \in A$ satisfy $q(a_1) = q(a_2) = b$ and define

$$q_3 = p \cup \{\langle a_1, b \rangle\}, \ \ q_4 = p \cup \{\langle a_2, b \rangle\}.$$

If $a_1 = a_2$, take $q_5 = q_3 = q_4$. If $a_1 \neq a_2$, take $q_5 = p \cup \{\langle a_1, b \rangle, \langle a_2, b \rangle\}$. ∎

Lemma 2. *If $\{U_j : j \in J\}$ is an antichain of **C**-closed sieves in P, then $\bigcup_{j \in J} U_j$ is **C**-closed and is accordingly the join of $\{U_j : j \in J\}$ in the frame H_C of **C**-closed sieves[28] in P.*

Proof. Let $S \in C(p)$ and suppose that $S \subseteq \bigcup_{j \in J} U_j$. We claim that $S \subseteq U_j$ for some (unique) $j \in J$. Given $q_1 \in S$, fix j so that $q_1 \in U_j$. For each $q_2 \in S$, there are, by

---

[27] For the definition of coverage see the Appendix.
[28] For the relevant definitions see the Appendix.

Lemma 1, $q_3, q_4, q_5 \in S$ such that $q_3 \subseteq q_1, q_4 \subseteq q_2$ and $q_3 \cup q_4 \subseteq q_5$. Since $q_3 \in S$ there is j' for which $q_3 \in U_{j'}$; since $U_{j'}$ is a sieve, it follows that $q_1 \in U_{j'}$. Hence $j = j'$ and so $q_3 \in U_j$. Hence $q_5 \in U_j$ and from this, by an argument similar to that establishing $q_3 \in U_{j'}$, it follows that $q_4 \in U_j$. We conclude that $q_2 \in U_j$. Since q_2 was an arbitrary member of S, it follows that $S \subseteq U_j$. Since U_j is C-closed and $S \in C(p)$, we infer that $p \in U_j$ and the Lemma is proved. ∎

Remark. It follows from Lemma 2 that H_C is connected. For if the top element P of H_{CS} is the join of an antichain $\{U_i : i \in I\}$ of elements of H_C, then by Lemma 2, $\bigcup_{i \in I} U_i = P$. Hence there is $i \in I$ for which $\emptyset \in U_i$; but then $U_i = P$.

Proposition 5. H_C *is completely* $\perp$-*distributive and so preserves exponentials.*

Proof. Let $\{U_{ij} : i \in I, j \in J\} \subseteq H_c$ be such that $\{U_{ij} : j \in J\}$ is an antichain for each $i \in I$. By Lemma 2, it suffices to show that

(*) $$\bigcap_{i \in I} \bigcup_{j \in J} U_{ij} \subseteq \bigcup_{f \in J^I} \bigcap_{i \in I} U_{if(i)}.$$

If $p \in \bigcap_{i \in I} \bigcup_{j \in J} U_{ij}$, then for each $i \in I$ there is a unique $j \in J$ such that $p \in U_{ij}$. We define $h \in J^I$ by taking $h(i)$ to be this unique j. Then $p \in \bigcap_{i \in I} U_{ih(i)}$, whence $p \in \bigcup_{f \in J^I} \bigcap_{i \in I} U_{if(i)}$ and (*) follows. ∎

Remark. It follows from Props 4 and 5 that H_C is locally connected.

Proposition 6 $\mathbf{\mathcal{B}}^{(H_C)} \models \widehat{B}$ *is a subquotient of* $\widehat{A}$.

Proof. By the Corollary to Prop. 1, it suffices to show that there is a subset $\{U_{ab} : a \in A, b \in B\} \subseteq H_c$ such that $U_{ab} \cap U_{ab'} = \emptyset$ for $b \neq b'$ and $\bigvee_{a \in A} U_{ab} = \top$ for

$b \in B$. Define $U_{ab} = \{p : \langle a, b \rangle \in p\}$. Then U_{ab} is C-closed. For suppose the cover S of p determined by $b' \in B$ is included in U_{ab}. We need to show that $p \in U_{ab}$. If $b' \in \text{ran}(p)$, then $p \in S \subseteq U_{ab}$. If $b' \notin \text{ran}(p)$, choose $a' \notin \text{dom}(p) \cup \{a\}$ and define $p' = p \cup \{\langle a', b' \rangle\}$. Then $p' \in S$, so that $p' \in U_{ab}$, from which it easily follows that $p \in U_{ab}$. Clearly $U_{ab} \cap U_{ab'} = \emptyset$ for $b \neq b'$.

To show finally that $\bigvee_{a \in A} U_{ab} = \top$, for $b \in B$, it suffices to show that P s the only C-closed sieve containing $\bigcup_{a \in A} U_{ab}$, and for this it suffices to show that for each $p \in P$ there is $S \in C(p)$ for which $S \subseteq \bigcup_{a \in A} U_{ab}$. Given $p \in P$, let S be the cover of p determined by b. If $q \in S$. then $q(a) = b$ for some $a \in A$, whence $q \in \bigcup_{a \in A} U_{ab}$. It follows that $S \subseteq \bigcup_{a \in A} U_{ab}$. ∎

Now in the preceding take $A = \mathbb{N}$ and $B = \mathbb{N}^{\mathbb{N}}$. Then by Propositions 5 and 6

$$\mathfrak{V}^{(H_C)} \vDash \widehat{\mathbb{N}}^{\widehat{\mathbb{N}}} = \widehat{\mathbb{N}^{\mathbb{N}}} \text{ is a subquotient of } \widehat{\mathbb{N}},$$

and so

$$\mathfrak{V}^{(H_C)} \vDash \mathbb{N}^{\mathbb{N}} \text{ is subcountable.}$$

The relative consistency with **IZF** of the subcountability of $\mathbb{N}^{\mathbb{N}}$ follows.

Remark. Suppose that H is a frame containing a triply-indexed subset $\{a_{mnp} : m, n, p \in \mathbb{N}\}$ satisfying the conditions:

(1) $\quad a_{mnp} \wedge a_{mnq} = \bot$ for $p \neq q$

(2) $\quad \bigwedge_{m} \bigwedge_{n} \bigvee_{p} a_{mnp} = \top$

(3) $\quad \bigwedge_{f \in \mathbb{N}^{\mathbb{N}}} \bigvee_{m} \bigwedge_{n} a_{mnf(n)} = \top$

Then H cannot be $\bot$-$(\mathbb{N}, \mathbb{N})$ distributive.

To see this, let $n \mapsto \tilde{n}$ be any bijection of $\mathbb{N}$ with itself lacking fixed points and define $b_{nm} \in H$ by $b_{mn} = a_{mm\tilde{n}}$. Then we have

$$b_{mn} \wedge b_{mp} = \bot \text{ for } n \neq p$$

And

$$\bigvee_n b_{mn} = \bigvee_n a_{mm\tilde{n}} = \bigvee_p a_{mnp} = \top,$$

so that

$$\bigwedge_m \bigvee_n b_{mn} = \top.$$

On the other hand

$$\bigvee_{f \in \mathbb{N}^{\mathbb{N}}} \bigwedge_m b_{mf(m)} = \bot.$$

To prove this, notice that it follows from (3) that $\bigvee_m a_{mnf(m)} = \top$, whence

$$\bigwedge_m b_{mf(m)} = \bigwedge_m a_{mm\widetilde{f(m)}} = \bigvee_m a_{mnf(m)} \wedge \bigwedge_m a_{mm\widetilde{f(m)}} \leq \bigvee_m (a_{mnf(m)} \wedge a_{mm\widetilde{f(m)}}) = \bot.$$

It follows that H is not $\bot$-$(\mathbb{N}, \mathbb{N})$ distributive.

This argument has an analogue in $V^{(H)}$. Define $\varphi \in V^{(H)}$ by

$$\mathrm{dom}(\varphi) = \{\overline{\langle m, n, p \rangle} : m, n, p \in \mathbb{N}\}$$

and

$$\varphi(\overline{\langle m, n, p \rangle}) = a_{mnp}.$$

Then conditions (1) and (2) abve imply that

$$\mathbf{B}^{(H)} \models \varphi : \widehat{\mathbb{N}} \times \widehat{\mathbb{N}} \to \widehat{\mathbb{N}}.$$

and $[\varphi(\widehat{m}, \widehat{n}) = \widehat{p}] = a_{mnp}$.

Now, in $\mathbf{B}^{(H)}$, let $\psi : \widehat{\mathbb{N}} \to \widehat{\mathbb{N}}^{\widehat{\mathbb{N}}}$ be defined so that $\psi(\widehat{m})(\widehat{n}) = \varphi(\widehat{m}, \widehat{n})$. It then follows from (3) that

(*) $\qquad\qquad\qquad \mathbf{B}^{(H)} \models \psi \text{ is a surjection of } \widehat{\mathbb{N}} \text{ onto } \widehat{\mathbb{N}^{\mathbb{N}}}$.

But the usual diagonal argument, carried out in $\mathbf{B}^{(H)}$, shows that

$$\mathbf{B}^{(H)} \models \text{there is no surjection of } \widehat{\mathbb{N}} \text{ onto } \widehat{\mathbb{N}}^{\widehat{\mathbb{N}}},$$

and hence, using (*),

$$\mathbf{B}^{(H)} \models \widehat{\mathbb{N}^{\mathbb{N}}} \neq \widehat{\mathbb{N}}^{\widehat{\mathbb{N}}}.$$

It now follows from Proposition 3 that H is not $\bot$-$(\mathbb{N}, \mathbb{N})$ distributive.

THE AXIOM OF CHOICE IN FRAME-VALUED EXTENSIONS

If I is a set, the *Axiom of Choice for I* is the assertion:

AC(I) *for any formula φ and any set A*

$$\forall i \in I \exists x \in A \; \varphi(i, x) \to \exists f \in A^I \forall i \in I \; \varphi(i, f(i)).$$

AC($\mathbb{N}$) is known as the *Countable Axiom of Choice*.

Proposition 7. *If H is refinable, then $\mathbf{B}^{(H)} \models \mathbf{AC}(\widehat{I})$ for every set I. In particular, $\mathbf{B}^{(H)} \models \mathbf{AC}(\widehat{\mathbb{N}})$, so that the Countable Axiom of Choice holds in $\mathbf{B}^{(H)}$.*

Proof. We have

$$[\forall i \in \hat{I} \exists x \in A\varphi(i,x)] = \bigwedge_{i \in I} \bigvee_{x \in \text{don}(A)} A(x) \wedge [\varphi(\hat{i},x)].$$

Since H is refinable, we may use **AC** to select, for each $i \in I$, a disjoint refinement $B_i = \{b_{ij} : j \in J_i\}$ of $\{A(x) \wedge [\varphi(\hat{i},x)] : x \in \text{dom}(A)\}$ with the same join as the latter. Again using **AC**, select for each $i \in I$, $j \in J_i$ an element $x_{ij} \in \text{dom}(A)$ for which $b_{ij} \leq A(x_{ij}) \wedge [\varphi(\hat{i}, x_{ij})]$. If we now define $f \in V^{(H)}$ by $\text{dom}(f) = \{<\hat{i}, x_{ij}>^{(H)} : i \in I, j \in J_i\}$ and $f(<\hat{i}, x_{ij}>^{(H)}) = b_{ij}$, a tedious but straightforward calculation shows that

$$\bigwedge_{i \in I} \bigvee_{x \in \text{don}(A)} A(x) \wedge [\varphi(\hat{i},x)] = \bigwedge_{i \in I} \bigvee_{j \in J_i} b_{ij} \leq [f : \hat{I} \to A \wedge \forall i \in \hat{I} \varphi(i, f(i))].$$

$\mathcal{B}^{(H)} \models \mathbf{AC}(\hat{I})$ follows immediately. ∎

A frame H is *countably generated* if it has a countable subset S such that every element of H is the join of elements of S. If this is the case, S is called a *countable set of generators for H*.

Proposition 8. *If H is countably generated and totally disconnected, it is refinable.*

Proof. Let $\{a_i : i \in I\}$ be an arbitrary subset of a countably generated, totally disconnected frame. We first show that there is a countable subset $I_0 \subseteq I$ such that $\bigvee_{i \in I} a_i = \bigvee_{i \in I_0} a_i$.

Let S be a countable set of generators for H, and for each $i \in I$ choose $S_i \subseteq S$ so that $\bigvee S_i = a_i$. Then $T = \bigcup_{i \in I} S_i$ is, as a subset of S, countable, and so can be presented as $\{t_n : n \in \mathbb{N}\}$. Moreover $\bigvee_{n \in \mathbb{N}} t_n = \bigvee T = \bigvee_{i \in I} a_i$. For each $n \in \mathbb{N}$ there is

$i_n \in I$ such that $t_n \leq a_{i_n}$. Then $I_0 = \{i_n : n \in \mathbb{N}\}$ is a countable subset of I and
$$\bigvee_{i \in I} a_i = \bigvee_{i \in I_0} a_i.$$

Now since H is totally disconnected, each a_i is the join of a set of complemented elements, and by the argument above, this set may be taken to be countable. For each $i \in I$ let $\{b_{in} : n \in \mathbb{N}\}$ be a (countable) set of complemented elements such that $a_i = \bigvee_{n \in \mathbb{N}} b_{in}$. Then

$$\bigvee_{i \in I} a_i = \bigvee_{i \in I_0} a_i = \bigvee_{i \in I_0} \bigvee_{n \in \mathbb{N}} b_{in}.$$

Let $\{c_n : n \in \mathbb{N}\}$ be an enumeration of the countable set $\{b_{in} : i \in I_0, n \in \mathbb{N}\}$. Then $\{c_n : n \in \mathbb{N}\}$ is a refinement of $\{a_i : i \in I\}$ with the same join as the latter. Now define $d_n \in H$, for each n, recursively by

$$d_0 = c_0, \qquad d_{n+1} = c_{n+1} \wedge (d_0 \vee \ldots \vee d_n)^*.$$

Then $\{d_n : n \in \mathbb{N}\}$ is an antichain, $d_n \leq c_n$ for each n, and $\bigvee_{n \in \mathbb{N}} d_m = \bigvee_{n \in \mathbb{N}} c_n$. It follows that $\{d_n : n \in \mathbb{N}\}$ is a disjoint refinement of $\{a_i : i \in I\}$ with the same join as the latter. The refinability of H follows. ∎

REAL NUMBERS AND REAL FUNCTIONS IN SPATIAL EXTENSIONS

In $\mathcal{B}^{(H)}$, the set of rational numbers may be identified with the H-set $\widehat{\mathbb{Q}}$, where $\mathbb{Q}$ is the usual set of rational numbers. Since $\mathcal{B}^{(H)}$ is a model of **IZ**, we can carry out within it the construction from the rationals of the set $\mathbb{R}_d$ of (Dedekind) real numbers as in Ch. 3. Let $\mathbb{R}_H$ be a strong core for the resulting H-valued set. The members of are naturally thought of as H-valued real numbers. More generally, given $h \in H$, an H-valued real number of degree h. is an element $r \in V^{(H)}$ for which $h \leq [\![r \in \mathbb{R}_d]\!]$.

Since each Dedekind real is a cut in the rationals, each H-valued real number r is a pair $<\mathbf{L},\mathbf{R}> \in V^{(H)}$ for which

$$[<L,R> \text{ is a cut in } \widehat{\mathbb{Q}}] = \top.$$

This condition translates into conditions on the truth values $[\![\hat{p} \in L]\!]$, $[\![\hat{q} \in R]\!]$ viz.,

0. $[\![L \cup R \subseteq \widehat{\mathbb{Q}}]\!] = \top$
1. $\bigvee_{p,q \in \mathbb{Q}} [\![\hat{p} \in L]\!] \wedge [\![\hat{q} \in R]\!] = \top$
2. $[\![\hat{p} \in L]\!] \wedge [\![\hat{p} \in R]\!] = \bot$
3. $[\![\hat{p} \in L]\!] = \bigvee_{q>p} [\![\hat{q} \in L]\!]$
4. $[\![\hat{p} \in R]\!] = \bigvee_{q<p} [\![\hat{q} \in R]\!]$
5. $[\![\hat{p} \in L]\!] \vee [\![\hat{q} \in R]\!] = \top$ for $p < q$.

Similar conditions may be written down for H-valued real numbers of degree h: these are left to the reader.

It is easy to check that, writing $\mathbb{R}$ for the "genuine" set of real numbers in V, we have $\mathscr{B}^{(H)} \vDash \widehat{\mathbb{R}} \subseteq \mathbb{R}_d$. In particular, for each $r \in \mathbb{R}$ we have $V^{(H)} \vDash \hat{r} \in \mathbb{R}_d$. Hence we may assume that $\hat{r} \in \mathbb{R}_H$.

$\mathbb{R}_H$ can be turned into an *ordered ring* by defining $+$, $\cdot$, $<$ as follows: for $r, s \in \mathbb{R}_H$,

$r \oplus s = $ unique element u of $\mathbb{R}_H$ such that $[\![u = r + s]\!] = \top$
$r \odot s = $ unique element u of $\mathbb{R}_H$ such that $[\![u = r.s]\!] = \top$
$r \prec s \leftrightarrow [\![r < s]\!] = \top$

With these definitions $\mathbb{R}_H$ is called the *ordered ring of H-valued real numbers*.

Now let X be a topological space. For brevity we shall write $\mathscr{B}^{(X)}$, $V^{(X)}$, $\mathbb{R}_X$ for $\mathscr{B}^{(\mathcal{O}(X))}$, $V^{(\mathcal{O}(X))}$, $\mathbb{R}_{\mathcal{O}(X)}$ respectively. Members of $\mathbb{R}_X$ will be called simply *real* numbers over X. Thus a real number over X may be identified as an element $r = \langle \mathbf{L}, \mathbf{R} \rangle$ of $V^{(X)}$ satisfying the following conditions:

$0_X.$ $[\mathbf{L} \cup \mathbf{R} \subseteq \widehat{\mathbb{Q}}] = X$

$1_X.$ $\bigcup\limits_{p,q \in \mathbb{Q}} [\widehat{p} \in \mathbf{L}] \cap [\widehat{q} \in \mathbf{R}] = X$

$2_X.$ $[\widehat{p} \in \mathbf{L}] \cap [\widehat{p} \in \mathbf{R}] = \varnothing$

$3_X.$ $[\widehat{p} \in \mathbf{L}] = \bigcup\limits_{q > p} [\widehat{q} \in \mathbf{L}]$

$4_X.$ $[\widehat{p} \in \mathbf{R}] = \bigcup\limits_{q < p} [\widehat{q} \in \mathbf{R}]$

$5_X.$ $[\widehat{p} \in \mathbf{L}] \cup [\widehat{q} \in \mathbf{R}] = X$ for $p < q$.

If $U \in \mathcal{O}(X)$ is an open subset of X, an $\mathcal{O}(X)$ - valued real number of degree U will be called a *real number over U*. Conditions analogous to $0_X - 5_X$ above can be formulated for real numbers over U.

We now prove

Proposition 9. *The ordered ring* $\mathbb{R}_X$ *of real numbers over X is isomorphic to the ordered ring* $\mathbf{C}(X, \mathbb{R})$ *of continuous real-valued functions on X.*

Proof. To obtain this isomorphism, start with a real number $r = \langle \mathbf{L}, \mathbf{R} \rangle$ over X. For each $t \in X$ define

$$\mathbf{L}_t = \{p \in \mathbb{Q} : t \in [\widehat{p} \in \mathbf{L}]\} \qquad \mathbf{R}_t = \{p \in \mathbb{Q} : t \in [\widehat{p} \in \mathbf{R}]\}$$

and $r_t = \langle \mathbf{L}_t, \mathbf{R}_t \rangle$. Then $r_t \in \mathbb{R}$ and the map $r^*\colon t \mapsto r_t$ is continuous and hence an element of $\mathbf{C}(X, \mathbb{R})$.

To show that $r_t \in \mathbb{R}$ (i.e. r_t is a Dedekind real), we check, for example, the condition

$$\forall p(p \in L_t \leftrightarrow \exists q \in L_t . q > p).$$

Using condition 3_X, we have for $t \in X$

$$p \in L_t \leftrightarrow t \in [\hat{p} \in L] \leftrightarrow t \in \bigcup_{q>p} [\hat{q} \in L] \leftrightarrow \exists q > p. \ t \in [\hat{q} \in L]$$
$$\leftrightarrow \exists q > p. \ q \in L_t \leftrightarrow \exists q \in L_t . q > p.$$

The other conditions are checked similarly.

To show that r^* is continuous, it suffices to show that the inverse image under r^* of each open interval $(p, \rightarrow)$ in $\mathbb{R}$ is open in X. This follows from the observation that, for $t \in X$, we have

$$r_t = \langle L_t, R_t \rangle \in (p, \rightarrow) \leftrightarrow p \in L_t \leftrightarrow t \in [\hat{p} \in L].$$

Accordingly the inverse image under r^* of $(p, \rightarrow)$ is $[\hat{p} \in L]$, which is open in X.
∎

The function r^* is said to be *correlated* with r.

Remark. For each "genuine" real number $r \in V$, $\hat{r}^*$ is the constant function on X with value r. In general, if $\mathbf{B}^{(H)} \models r \in \hat{\mathbb{R}}$, then r^* is *locally constant*, that is, each point of X gas a neighbourhood on which r^* is constant. Note that, if X is connected, then each locally constant function on X is constant.

Conversely, given $f \in C(X, \mathbb{R})$, define $\mathbf{L}_f, \mathbf{R}_f \in V^{(X)}$ by

$$\text{dom}(\mathbf{L}_f) = \text{dom}(\mathbf{R}_f) = \{\hat{p} : p \in \mathbb{Q}\},$$

with

$$\mathbf{L}_f(\hat{p}) = f^{-1}((p, \rightarrow)), \quad \mathbf{R}_f(\hat{p}) = f^{-1}((\leftarrow, p)).$$

We claim that

$$\mathfrak{B}^{(H)} \models \langle \mathbf{L}_f, \mathbf{R}_f \rangle \in \mathbb{R}_d.$$

We verify conditions 3_X and 5_X, leaving the rest to the reader. First note that

$$[\hat{p} \in \mathbf{L}_f] = f^{-1}((p, \rightarrow)).$$

Ad 3_X:

$$[\hat{p} \in \mathbf{L}_f] = f^{-1}((p, \rightarrow)) = f^{-1}(\bigcup_{q>p}(q, \rightarrow)) = \bigcup_{q>p} f^{-1}((q, \rightarrow)) = \bigcup_{q>p} [\hat{q} \in \mathbf{L}_f].$$

Ad 5_X: For $p < q$, we have

$$[\hat{p} \in \mathbf{L}_f] \cup [\hat{q} \in \mathbf{R}_f] = f^{-1}(p, \rightarrow) \cup f^{-1}(\leftarrow, q) = f^{-1}(\mathbb{R}) = X.$$

We define $\overline{f}$ to be the unique element r of $\mathbb{R}_X$ for which $[r = \langle \mathbf{L}_f, \mathbf{R}_f \rangle] = X$. $\overline{f}$ is called the real number over X *correlated* with f.

If f is locally constant, it is easy to check that $\mathfrak{B}^{(H)} \models \overline{f} \in \hat{\mathbb{R}}$.

We next show that the maps $r \mapsto r^*$ and $f \mapsto \overline{f}$ are mutually inverse, i.e. $\overline{r^*} = r$ and $(\overline{f})^* = f$.

For the first assertion, we note that, for $t \in X$,

$$t \in [\hat{p} \in \mathbf{L}_{r^*}] \leftrightarrow t \in (r^*)^{-1}(p, \rightarrow) \leftrightarrow r^*(t) \in (p, \rightarrow)$$
$$\leftrightarrow r_t > p \leftrightarrow p \in \mathbf{L}_t \leftrightarrow t \in [\hat{p} \in \mathbf{L}].$$

It follows that $[\hat{p} \in \mathbf{L}_{r^*}] = [\hat{p} \in \mathbf{L}]$, whence $[\mathbf{L}_{r^*} = \mathbf{L}] = X$. Similarly $[\mathbf{R}_{r^*} = \mathbf{R}] = X$, whence $[\overline{r^*} = r] = X$, so that $\overline{r^*} = r$.

89

For the second assertion, note that $(\overline{f})*(t) = <(L_f)_t, (R_f)_t>$. So if $f(t) = $ <L, R>, then

$$p \in (L_f)_t \leftrightarrow t \in [\hat{p} \in L_f] \leftrightarrow t \in f^{-1}((p, \rightarrow)) \leftrightarrow f(t) > p \leftrightarrow p \in L.$$

Accordingly $(L_f)_t = L$. Similarly $(R_f)_t = R$, whence $(\overline{f})*(t) = f(t)$. Since this holds for arbitrary $t \in X$, it follows that $(\overline{f})* = f$.

We claim finally that the map $r \mapsto r*$ is an isomorphism of $\mathbb{R}_X$ with $C(X, \mathbb{R})$. To establish this it suffices to show that

(*) $\qquad r \prec s \leftrightarrow r* < s* \qquad (r \oplus s)* = r* + s* \qquad (r \odot s)* = r* \cdot s*$

The first of these assertions is an immediate consequence of the fact that $[r < s] = \{t : r*(t) < s*(t)\}$. This latter is proved as follows: Let $r = $ <L, R>, $s = $ <L', R'>. Then we have

$$t \in [r < s] \leftrightarrow t \in [\exists p[p \in R \wedge p \in L'] \leftrightarrow t \in \bigcup_p [\hat{p} \in R \wedge p \in L']$$
$$\leftrightarrow \exists p[t \in [\hat{p} \in R] \cap [\hat{p} \in L'] \leftrightarrow \exists p[p \in R_t \wedge p \in R_t]$$
$$\leftrightarrow r_t < s_t \leftrightarrow r*(t) < s*(t).$$

The proofs of the remaining assertions in (*) are left to the reader. ∎

The upshot of Proposition 9 is that real numbers over X can be regarded as real numbers *varying continuously* over X.

For the record, we also note:

$$[\overline{f} = \overline{g}] = \text{In}\{t : f(t) = g(t)\}.\ [29]$$

This follows from:

---

[29] Here In denotes the interior operation in a topological space.

$$\begin{aligned}
[\overline{f = g}] = [\mathsf{L}_f = \mathsf{L}_g] &\cap [\mathsf{R}_f = \mathsf{R}_g] \\
&= \operatorname{In}\bigcap_p [\hat{p} \in \mathsf{L}_f] \Leftrightarrow [\hat{p} \in \mathsf{L}_g] \cap \operatorname{In}\bigcap_p [\hat{p} \in \mathsf{R}_f] \Leftrightarrow [\hat{p} \in \mathsf{R}_g] \\
&= \operatorname{In}\bigcap_p (f^{-1}(p, \to) \Leftrightarrow g^{-1}(p, \to)) \cap \operatorname{In}\bigcap_p (f^{-1}(\leftarrow, p) \Leftrightarrow g^{-1}(\leftarrow, p)) \\
&= \operatorname{In}\bigcap_p [(f^{-1}(p, \to) \Leftrightarrow g^{-1}(p, \to)) \cap (f^{-1}(\leftarrow, p) \Leftrightarrow g^{-1}(\leftarrow, p))] \\
&= \operatorname{In}\{t : \forall p [(f(t) > p \leftrightarrow g(t) > p) \wedge (f(t) < p \leftrightarrow g(t) < p)] \\
&= \operatorname{In}\{t : f(t) = g(t)\}. \blacksquare
\end{aligned}$$

In a similar way, one shows that, for real numbers r, s over X, we have

$$[r = s] = \operatorname{In}\{t : r * (t) = s * (t)\}.$$

and

$$[r \leq s] = \operatorname{In}\{t : r * (t) \leq s * (t)\}.$$

These arguments can easily be extended to establish, for any open set U in X, a natural correspondence, with analogous properties, between real numbers over U and continuous real-valued functions on U. Thus, writing $C(U, \mathbb{R})$ for the set of real-valued continuous functions on U, real numbers over U *correspond* to elements of $C(U, \mathbb{R})$. Under this correspondence locally constant functions on U are associated with real numbers r over U for which $U \subseteq]r \in \hat{\mathbb{R}}]$.

A *real function over* X is an element $\varphi \in V^{(X)}$ such that

$$\mathcal{B}^{(X)} \vDash \operatorname{Fun}(\varphi) \wedge \operatorname{dom}(\varphi) = \mathbb{R}_d \wedge \operatorname{ran}(\varphi) \subseteq \mathbb{R}_d$$

Since real numbers over X correspond to elements of $C(X, \mathbb{R})$, real functions over X should be correlated with certain *operators* on $C(X, \mathbb{R})$, that is, maps $\Phi: C(X, \mathbb{R}) \to C(X, \mathbb{R})$. We now set about identifying these operators.

An operator Φ on $C(X, \mathbb{R})$ is said to be

- *near-local* if, for any $f, g \in C(X, \mathbb{R})$,
$$\text{In}\{t: f(t) = g(t)\} \subseteq \{t: \Phi(f)(t) = \Phi(g)(t)\},$$
or equivalently, if
$$\text{In}\{t: f(t) = g(t)\} \subseteq \text{In}\{t: \Phi(f)(t) = \Phi(g)(t)\},$$

- *local* if, for any $f, g \in C(X, \mathbb{R})$,
$$\{t: f(t) = g(t)\} \subseteq \{t: \Phi(f)(t) = \Phi(g)(t)\}.$$

Clearly any local operator is near-local. In general, the converse is false, but we shall later show that, for metric spaces, every near-local operator is local.

We next show that real functions over X are correlated with near-local operators on $C(X, \mathbb{R})$.

Given a real function φ over X, define the operator Φ on $C(X, \mathbb{R})$ by $\Phi(f) = \varphi(\overline{f})^*$ for $f \in C(X, \mathbb{R})$. Φ is the operator *correlated* with φ.

We claim that Φ is near-local. To establish this, note that
$$\text{In}\{t : f(t) = g(t)\} = [\overline{f} = \overline{g}] \subseteq [\varphi(\overline{f}) = \varphi(\overline{g})] = \text{In}\{t : \varphi(\overline{f})^*(t) = \varphi(\overline{g})^*(t)\}$$
$$= \text{In}\{t : \Phi(f)(t) = \Phi(g)(t)\}.$$

Now suppose given a *local* operator Δ on $C(X, \mathbb{R})$. We define the function $D: X \times \mathbb{R} \to \mathbb{R}$ by the stipulation:

$D(t, a) = b$ iff for some $f \in C(X, \mathbb{R})$, $f(t) = a$ and $\Delta(f)(t) = b$.

D is called the function on $X \times \mathbb{R}$ *correlated* with Δ. Clearly D satisfies

$$D(t, f(t)) = \Delta(f)(t)$$

for arbitrary $t \in X$, $f \in C(X, \mathbb{R})$.

Let us call a function $F: X \times \mathbb{R} \to \mathbb{R}$ *localizable* if, for some local operator Δ on $C(X, \mathbb{R})$, we have $F(t, f(t)) = \Delta(f)(t)$ for arbitrary $t \in X$, $f \in C(X, \mathbb{R})$. Local operators on $C(X, \mathbb{R})$ are thus correlated with localizable functions on $X \times \mathbb{R}$.

PROPERTIES OF THE SET OF REAL NUMBERS OVER $\mathbb{R}$

We now focus attention on the case in which the space X is the space $\mathbb{R}$ of real numbers.

To begin with, let i, a be the real numbers over $\mathbb{R}$ correlated with the *identity* function and the *absolute value* function, respectively, on $\mathbb{R}$. Then we have

- $\mathcal{B}^{(\mathbb{R})} \not\vdash i < 0 \vee i = 0 \vee i > 0$. That is, the *law of trichotomy for $\mathbb{R}_d$ is not affirmed in* $\mathcal{B}^{(\mathbb{R})}$.

This follows from the observation that

$$[\![i < 0 \vee i = 0 \vee i > 0]\!] = (\leftarrow, 0) \cup \mathrm{In}\{0\} \cup (0, \to) = (\leftarrow, 0) \cup (0, \to)$$
$$= \mathbb{R} - \{0\} \neq \mathbb{R}.$$

Similarly, one shows that

93

- $\mathcal{B}^{(\mathbb{R})} \not\Vdash i \leq 0 \vee 0 \leq i$.
- $\mathcal{B}^{(\mathbb{R})} \not\Vdash i \leq a \leftrightarrow (i < a \vee i = a)$

and

- $\mathcal{B}^{(\mathbb{R})} \not\Vdash i = a \vee i \neq a$, so that, $\mathcal{B}^{(\mathbb{R})} \not\Vdash \mathbb{R}_d$ is discrete.
- While $\mathcal{B}^{(\mathbb{R})} \vDash \iota \neq 0$, $\mathcal{B}^{(\mathbb{R})} \not\Vdash \iota$ is invertible. Thus $\mathcal{B}^{(\mathbb{R})} \not\Vdash \mathbb{R}_d$ is a field.

To see this, first observe that that $\llbracket i = 0 \rrbracket = \text{In}\{0\} = \emptyset$. Also

$$\llbracket i \text{ is invertible} \rrbracket = \llbracket \exists x(ix = 1) \rrbracket = \bigcup_{U \in \mathcal{O}(\mathbb{R})} \bigcup_{f \in C(U, \mathbb{R})} \text{In}\{t : tf(t) = 1\} \not\ni 0$$

Hence $\llbracket i \text{ is invertible} \rrbracket \neq \mathbb{R}$.

- $\mathcal{B}^{(\mathbb{R})} \vDash i$ is not the limit of a Cauchy sequence of rationals. It follows that both $\mathbb{R}_c = \mathbb{R}_d$ and **AC(N)** are false in $\mathcal{B}^{(\mathbb{R})}$.

We sketch a proof of this. It is required to show that

(*) $\quad \llbracket \exists u \in \widehat{\mathbb{Q}}^{\widehat{\mathbb{N}}} [u \text{ is a Cauchy sequence of rational numbers converging to } i] \rrbracket = \emptyset$

Since $\mathbb{R}$ is locally connected, $\mathcal{B}^{(\mathbb{R})} \vDash \widehat{\mathbb{Q}}^{\widehat{\mathbb{N}}} = \widehat{\mathbb{Q}^{\mathbb{N}}}$, so (*) is equivalent to

for all $s \in \mathbb{Q}^{\mathbb{N}}$, $\llbracket \hat{s} \text{ is a Cauchy sequence of rational numbers converging to } i \rrbracket = \emptyset$

Accordingly it will enough to show that, for any $U \in \mathcal{O}(\mathbb{R}), s \in \mathbb{Q}^{\mathbb{N}}$

(**) $\quad U \subseteq \llbracket \hat{s} \text{ is a Cauchy sequence of rational numbers converging to } \iota \rrbracket \rightarrow U = \emptyset$.

Again, because $\mathbb{R}$ is locally connected, it suffices to prove (**) for *connected* U.

So suppose U connected and

$$U \subseteq [\hat{s} \text{ is a Cauchy sequence of rational numbers converging to } i].$$

Then for each n, $U \subseteq [\hat{s_n} \in \hat{\mathbb{Q}}]$ and so $\hat{s_n}$ corresponds to a locally constant rational-valued function in $C(U, \mathbb{R})$. Since U is connected, this latter function is constant on U; let p_n be that constant (rational) value. The sequence $\langle p_n : n \in \mathbb{N} \rangle$ is then Cauchy; so it converges to some unique $t_0 \in \mathbb{R}$. We then have

$$U \subseteq [\hat{s} \text{ is a Cauchy sequence of rational numbers converging to } i]$$
$$= [\forall q \in \hat{\mathbb{Q}} \exists n \in \hat{\mathbb{N}} \forall m \geq n (|\hat{s_m} - t| < q]$$
$$= \text{In} \bigcap_q \bigcup_n \text{In} \bigcap_{m \geq n} \{t \in U : |p_n - t| < q\}$$
$$\subseteq \text{In}\{t: \langle p_n \rangle \text{ converges to } t\}$$
$$= \text{In}\{t_0\}$$
$$= \emptyset.$$

Accordingly $U = \emptyset$ and (**) follows.

PROPERTIES OF THE SET OF REAL NUMBERS OVER BAIRE SPACE

If we endow $\mathbb{N}$ with the discrete topology, the set $\mathbb{N}^\mathbb{N}$ endowed with the product topology will be written $\widetilde{N}$ and called *Baire space*. $\widetilde{N}$ is totally disconnected[30] and has a countable base consisting of clopen sets. We proceed to establish various properties of the spatial extension $\mathcal{B}^{(\widetilde{N})}$. Our principal task will be to show that, in $\mathcal{B}^{(\widetilde{N})}$, Brouwer's Principle holds, that is, in $\mathcal{B}^{(\widetilde{N})}$, every function from reals to reals is continuous.

We first note that, since $\mathcal{O}(\widetilde{N})$ is countably generated and totally disconnected, by Proposition 8 it is refinable, and hence, by Proposition 7, the Countable

---

[30] $\widetilde{N}$ can be shown to be homeomorphic to the space of irrational numbers.

Axiom of Choice holds in $\mathcal{B}^{(\widetilde{N})}$ Thus, in $\mathcal{B}^{(\widetilde{N})}$), every Dedekind real is the limit of a Cauchy sequence of rationals, and it follows that, in $\mathcal{B}^{(\widetilde{N})}$), *the Cauchy reals and the Dedekind reals coincide.*

We now proceed to outline the strategy (due to Scott [1970]) for showing that, in $\mathcal{B}^{(\widetilde{N})}$) , every function from reals to reals is continuous.

Step 1. Show that, when X is a metric space, every near-local operator on $C(X, \mathbb{R})$ is local.

Step 2. Show that every localizable function on $\widetilde{N} \times \mathbb{R}$ is continuous.

Step 3. Infer that real functions over $\widetilde{\Lambda}$ are correlated with continuous localizable functions on $\widetilde{N} \times \mathbb{R}$

Step 4. Show that each real function over $\widetilde{N}$ correlated with a continuous localizable function on $\widetilde{N} \times \mathbb{R}$ is continuous in $\mathcal{B}^{(\widetilde{N})}$

Step 5. Conclude that every real function over $\widetilde{\Lambda}$ is continuous in $\mathcal{B}^{(\widetilde{N})}$) .

The topological details for carrying out Steps 1 and 2 – which are somewhat intricate and are omitted here - may be found in Scott [1970]. Step 3 then follows accordingly.

Now for Step 4. First we note that since $\widetilde{N}$ is totally disconnected, the Definite Elements Lemma applies to $\mathcal{B}^{(\widetilde{N})}$ This should be borne in mind in the course of that argument that follows.

Now let φ be a real function over $\widetilde{N}$, correlated with the continuous localizable function $F: \widetilde{N} \times \mathbb{R} \to \mathbb{R}$ To show that φ is continuous in $\mathcal{B}^{(\widetilde{N})}$,we need to show that the sentence – which we shall denote by (¶) -

$$\forall p \forall q > p \vee e > 0 \exists d > 0 \forall x \in [p,q] \forall y \in [p,q][\,|x-y|<d \to |\varphi(x)-\varphi(y)|<e]$$

holds[31] in $\mathfrak{B}^{(\widetilde{N})}$. (Here we have used p, q, e, d as rational number variables, and $[p, q]$ denotes the closed interval in $\mathbb{R}$.) For this it suffices to show that, for any rationals p, q with $p < q$ and any positive rational e,

(*) $\quad \widetilde{N} \;=\; [\exists d > 0 \forall x \in [p,q] \forall y \in [p,q][\,|x-y|<d \to |\varphi(x)-\varphi(y)|<e]].$

Using the Definite Element Lemma and the correlation between real numbers over $\mathcal{N}$ and elements of $C(\widetilde{N}, \mathbb{R})$, proving (*) amounts to showing that $\widetilde{N}$ is identical with the set

$$\bigcup_{d>0} \operatorname{In} \bigcap_{f,g \in C(\widetilde{N}, \mathbb{R})} \operatorname{In}\{t : [f(t), g(t)] \in [p,q] \wedge |f(t) - g(t)| < d$$
$$\to |F(t, f(t)) - F(t, g(t))| < e\}.$$

We write S for this set.

This is proved by introducing the function $e: \mathcal{N} \times (0, \to) \to \mathbb{R}$ defined by

$$e(t, d) = \sup\{|F(t, x) - F(t, y)| : x, y \in [p,q], |x-y| < d\}.$$

The function $|F(t, x) - F(t, y)|$ is continuous in the variables t, x, y and the supremum is taken over a compact subset of $\mathbb{R} \times \mathbb{R}$. It follows that, for fixed d, $e(t, d)$ is a well-defined, continuous function of t. Now for fixed t the real function $F(t, x)$ is uniformly continuous for $x \in [p, q]$, and so $e(t, d) \to 0$ as $d \to 0$. Accordingly, given $e > 0$ and $t_0 \in \mathcal{N}$, we may choose $d > 0$ so that $e(t_0, d) < e$. Since e is continuous there is a neighbourhood U of t_0 in $\widetilde{N}$ such that $e(t, d) < e$ for all $t \in U$. It is now easily seen that U is included in S, and so the latter coincides with $\mathcal{N}$.

This completes Step 5 and we conclude that, in $\mathfrak{B}^{(\widetilde{N})}$, every function from reals to reals is continuous.

---

[31] In fact this sentence asserts that φ is *uniformly continuous on closed intervals*.

Remark. Call a set A *cohesive* if, whenever $A = U \cup V$ with $U \cap V = \emptyset$ then $U = A$ or $V = A$. A is cohesive iff every 2-valued function on A is constant, and it is not hard to show that this is equivalent to the condition that every $\mathbb{N}$-valued function is constant.

It follows from the truth of Brouwer's Principle in $\mathbf{B}^{(\tilde{N})}$ that, in $\mathbf{B}^{(\tilde{N})}$, $\mathbb{R}$ is cohesive. To prove this we show that from (¶) above it follows (in **IZ**) that any function $\varphi : \mathbb{R} \to 2$ is constant. Thus let p, q be rational numbers with $p < q$, and take $e = 1$ in (¶). We get a rational $d > 0$ satisfying

(†) $$\forall xy \in [p,q][\,|x - y| < d \to |f(x) - f(y)| < 1 \to f(x) = f(y)].$$

Let d' be a rational such that $0 < d' < d$. Let n be the least integer such that $q - p < nd'$, let $p_0, \ldots, p_n$ be defined by $p_0 = p$, $p_{i+1} = p_i + d'$ and define $K_i = [p_i, p_{i+1}] \cap [p, q]$. Then

$$[p,q] = \bigcup_{i=0}^{n-1} K_i.$$

A straightforward inductive argument, using (†), now shows that φ has constant value $\varphi(p)$ on each K_i and so also on $[p, q]$. Since p and q were arbitrary with $p < q$, φ is constant on the whole of $\mathbb{R}$.

THE INDEPENDENCE OF THE FUNDAMENTAL THEOREM OF ALGEBRA FROM IZF

The *Fundamental Theorem of Algebra* (**FTA**) asserts that the field $\mathbb{C}$ of complex numbers is algebraically closed, i.e. that every polynomial over $\mathbb{C}$ has a zero in $\mathbb{C}$. While **FTA** is provable in **ZF**, we shall establish its unprovability in **IZF** by showing that it is *false* in the spatial extension $\mathbf{B}^{(\mathbb{C})}$, where now $\mathbb{C}$ is the space of complex numbers with the usual topology.

In the same way as for $\mathbb{R}$, one shows the complex numbers over any open subset U of a topological space X are correlated with continuous functions in $C(U, \mathbb{C})$. This holds in particular when X is $\mathbb{C}$ itself. Write ι for the complex number over $\mathbb{C}$ correlated with the identity function on $\mathbb{C}$.

Now consider the polynomial $p(x) = x^2 - \iota$ in $\mathcal{B}(\mathbb{C})$.

For $u \in V^{(\mathbb{C})}$ and $U \in \mathcal{A}(\mathbb{C})$, if $U \subseteq [u \in \mathbb{C}]$, then u may be considered a complex number over U and accordingly is correlated with a continuous function $f : U \to \mathbb{C}$. In that case $[p(u) = 0] = [u^2 = \iota] = \text{In}\{t \in U : f(t)^2 = t\}$.

If $0 \in [p(u) = 0]$, then there is a neighbourhood V of 0 such that $V \subseteq \text{In}\{t \in U : f(t)^2 = t\}$. Let $q > 0$ be a rational number such that the circles C, C' about the origin with radii q, q^2 are both contained in V. Then since $f(t)^2 = t$ in V, the restriction of f to C' is a section of the squaring function $t \mapsto t^2 : C \to C'$. Thus this restriction would have to be a homeomorphism of C' to half of C. But this is impossible since any circle, but no half-circle, remains connected when a single (interior) point is removed.

Thus $0 \notin [p(u) = 0]$. Since this holds for arbitrary u, it follows that $0 \notin [\exists x(p(x) = 0)]$. Similarly, for any $a \in \mathbb{C}$, $a \notin [\exists x(p(x) + a = 0)]$. From this we deduce that $[\mathbb{C}$ is algebraically closed$] = \emptyset$, i.e., in $\mathcal{B}(\mathbb{C})$, $\mathbb{C}$ is not algebraically closed.

Appendix
Heyting Algebras, Frames and Intuitionistic Logic

LATTICES

A *lattice* is a (nonempty) partially ordered set L with partial ordering $\leq$ in which each two-element subset $\{x, y\}$ has a supremum or *join* – denoted by $x \vee y$ – and an infimum or *meet* – denoted by $x \wedge y$. A *top* (*bottom*) element of a lattice L is an element, denoted by $\top$ ($\bot$) such that $x \leq \top$ ($\bot \leq x$) for all $x \in L$. A lattice with top and bottom elements is called *bounded*. A lattice is *trivial* if it contains just one element, or equivalently, if in it $\bot = \top$. A *sublattice* of a bounded lattice L is a subset of L containing $\top$ and $\bot$ and closed under L's meet and join operations. It is easy to show that the following hold in any bounded lattice:

$$x \vee \bot = x, \quad x \wedge \top = x,$$
$$x \vee x = x, \quad x \wedge x = x,$$
$$x \vee y = y \vee x, \quad x \wedge y = y \wedge x,$$
$$x \vee (y \vee z) = (x \vee y) \vee z, \quad x \wedge (y \wedge z) = (x \wedge y) \wedge z,$$
$$(x \vee y) \wedge y = y, \quad (x \wedge y) \vee y = y$$

Conversely, suppose that $(L, \vee, \wedge, \bot, \top)$ is an algebraic structure, with $\vee, \wedge$ binary operations, in which the above equations hold, and define the relation $\leq$ on L by $x \leq y$ iff $x \vee y = y$. It is then easily shown that $(L, \leq)$ is a bounded lattice in which $\vee$ and $\wedge$ are, respectively, the join and meet operations, and 1 and 0 the top and bottom elements. This is the *equational characterization* of lattices.

Examples. (i) Any linearly ordered set is a lattice; clearly in this case we have $x \wedge y = \min(x, y)$ and $x \vee y = \max(x, y)$.

(ii) For any set A, the power set PA is a lattice under the partial ordering of set inclusion. In this lattice $X \vee Y = X \cup Y$ and $X \wedge Y = X \cap Y$. A sublattice of a power set lattice is called a *lattice of sets*.

(iii) If X is a topological space, the families $\mathcal{O}(X)$ and $\mathcal{C}(X)$ of open sets and closed sets, respectively, in X each form a lattice under the partial ordering of set inclusion. In these lattices $\vee$ and $\wedge$ are the same as in example (ii).

A lattice is said to be *distributive* if the following identities are satisfied:

$$x \wedge (y \vee z) = (x \wedge y) \vee (x \wedge z), \quad x \vee (y \wedge z) = (x \vee y) \wedge (x \vee z).$$

In the sequel by the term "distributive lattice" we shall understand "bounded distributive lattice." An easy inductive argument shows that any nonempty finite subset $\{x_1, \ldots, x_n\}$ of a lattice has a supremum and an infimum: these are denoted respectively by $x_1 \vee \ldots \vee x_n$, $x_1 \wedge \ldots \wedge x_n$. An arbitrary subset of a lattice need not have an infimum or a supremum: for example, the set of even integers in the totally ordered lattice of integers has neither. If a subset X of a given lattice does possess an infimum, or *meet*, it is denoted by $\bigwedge X$; if the subset possesses a supremum, or *join*, it is denoted by $\bigvee X$. When X is presented in the form $X = \{t(x): \varphi(x)\}$, $\bigwedge X$ and $\bigvee X$, if they exist, are written respectively $\bigwedge_{\varphi(x)} t(x)$ and $\bigvee_{\varphi(x)} t(x)$. When X is given in the form of an indexed set $\{x_i: i \in I\}$, its join and meet, if they exist, are written respectively $\bigvee_{i \in I} x_i$ and $\bigwedge_{i \in I} x_i$.

A lattice is *complete* if every subset has an infimum and a supremum. The meet and join of the empty subset of a complete lattice are, respectively, its top and bottom elements. It is a curious fact that, for a lattice to be complete, it suffices that every subset have a supremum, or every subset an infimum. For the supremum (infimum), if it exists, of the set of lower (upper)[32] bounds of a given subset X is easily seen to be the infimum (supremum) of X.

Examples. (i) The power set lattice $\mathbf{P}A$ of a set A is a complete lattice in which joins and meets coincide with set-theoretic unions and intersections respectively.

---

[32] Here by a *lower (upper) bound* of a subset X of a partially ordered set P we mean an element $p \in P$ for which $a \leq x$ ($x \leq a$) for every $x \in X$.

(ii) The lattices $\mathcal{O}(X)$ and $\mathcal{C}(X)$ of open sets and closed sets of a topological space are both complete. In $\mathcal{O}(X)$ the join and meet of a subfamily $\{U_i : i \in I\}$ are given by

$$\bigvee_{i \in I} U_i = \bigcup_{i \in I} U_i \qquad \bigwedge_{i \in I} U_i = \operatorname{In} \bigcap_{i \in I} U_i .$$

In $\mathcal{C}(X)$ the join and meet of a subfamily $\{A_i : i \in I\}$ are given by

$$\bigvee_{i \in I} A_i = \overline{\bigcup_{i \in I} A_i} \qquad \bigwedge_{i \in I} A_i = \bigcap_{i \in I} A_i .$$

Here $\operatorname{In} A$ and $\overline{A}$ denote the interior and closure, respectively, of a subset A of a topological space.

HEYTING AND BOOLEAN ALGEBRAS

A *Heyting algebra* is a bounded lattice $(H, \leq)$ such that, for any pair of elements $x, y \in H$, the set of $z \in H$ satisfying $z \wedge x \leq y$ has a *largest* element. This element, which is uniquely determined by x and y, is denoted by $x \Rightarrow y$: thus $x \Rightarrow y$ is characterized by the following condition: for all $z \in H$,

$$z \leq x \Rightarrow y \text{ if and only if } z \wedge x \leq y.$$

The binary operation on a Heyting algebra which sends each pair of elements x, y to the element $x \Rightarrow y$ is called *implication*; the operation which sends each element x to the element $x^* = x \Rightarrow \bot$ is called *pseudocomplementation*. We also define the operation $\Leftrightarrow$ of *equivalence* by $x \Leftrightarrow y = (x \Rightarrow y) \wedge (y \Rightarrow x)$. These operations are easily shown to satisfy:

$$x \Rightarrow (y \Rightarrow z) = (x \wedge y) \Rightarrow z, \quad x \Rightarrow y = \top \leftrightarrow x \leq y, \quad x \Leftrightarrow y = \top \leftrightarrow x = y,$$
$$y \leq z \to (x \Rightarrow y) \leq (x \Rightarrow z), \quad x \wedge (x \Rightarrow y) \leq y$$
$$y \leq x^* \leftrightarrow y \wedge x = \bot \leftrightarrow x \leq y^*, \quad x \leq x^{**}, \quad x^{***} = x^*, \quad (x \vee y)^* = x^* \wedge y^*.$$

To establish the last of these, observe that

$$z \leq (x \vee y)^* \leftrightarrow z \wedge (x \vee y) = \bot$$
$$\leftrightarrow (z \wedge x) \vee (z \wedge y) = \bot$$
$$\leftrightarrow z \wedge x = \bot \ \& \ z \wedge y = \bot$$
$$\leftrightarrow z \leq x^* \ \& \ z \leq y^*$$
$$\leftrightarrow z \leq x^* \wedge y^*.$$

Any Heyting algebra is a distributive lattice. To see this, calculate as follows for arbitrary elements x, y, z, u:

$$x \wedge (y \vee z) \leq u \leftrightarrow y \vee z \leq x \Rightarrow u$$
$$\leftrightarrow y \leq (x \Rightarrow u) \ \& \ z \leq (x \Rightarrow u)$$
$$\leftrightarrow x \wedge y \leq u \ \& \ x \wedge z \leq u$$
$$\leftrightarrow (x \vee y) \wedge (x \vee z) \leq u.$$

Any linearly ordered set with top and bottom elements is a Heyting algebra in which

$$x \Rightarrow y = \top \text{ if } x \leq y \qquad x \Rightarrow y = y \text{ if } y < x.$$

A basic fact about *complete* Heyting algebras is that the following identity holds in them:

(*) $$x \wedge \bigvee_{i \in I} y_i = \bigvee_{i \in I} x \wedge y_i$$

And conversely, in any complete lattice satisfying (*), defining the operation $\Rightarrow$ by $x \Rightarrow y = \bigvee \{z : z \wedge x \leq y\}$ turns it into a Heyting algebra.

To prove this, we observe that in any complete Heyting algebra,

$$\begin{aligned}
x \wedge \bigvee_{i\in I} y_i \leq z &\leftrightarrow \bigvee_{i\in I} y_i \leq x \Rightarrow z \\
&\leftrightarrow y_i \leq x \Rightarrow z, \text{ all } i \\
&\leftrightarrow y_i \wedge x \leq z, \text{ all } i \\
&\leftrightarrow \bigvee_{i\in I} x \wedge y_i \leq z.
\end{aligned}$$

Conversely, if (*) is satisfied and $x \Rightarrow y$ is defined as above, then

$$(x \Rightarrow y) \wedge x \leq \bigvee\{z : z \wedge x \leq y\} \wedge x = \bigvee\{z \wedge x : z \wedge x \leq y\} \leq y.$$

So $z \leq x \Rightarrow y \rightarrow z \wedge x \leq (x \Rightarrow y) \wedge x \leq y$. The reverse inequality is an immediate consequence of the definition.

In view of this result a complete Heyting algebra may also be defined to be a complete lattice satisfying (*). Complete Heyting algebras are known as *frames*.

If X is a topological space, then the complete lattice $\mathcal{O}(X)$ of open sets in X is a Heyting algebra. In $\mathcal{O}(X)$ meet and join are just set-theoretic intersection and union, while the implication and pseudocomplementation operations are given by $U \Rightarrow V = \text{In}((X - U) \cup V)$ and $U^* = \text{In}(X - U)$.

Let L be a bounded lattice. A *complement* for an element $a \in L$ is an element $b \in L$ satisfying $a \vee b = \top$ and $a \wedge b = \bot$. In general, an element of a lattice may have more than one complement, or none at all. However, in a *distributive* lattice an element can have at most one complement. For if b, b' are complements of an element a of a distributive lattice, then $a \vee b = a \vee b' = \top$ and $a \wedge b = a \wedge b' = \bot$. From this we deduce

$$b = b \vee \bot = b \vee (a \wedge b') = (b \vee a) \wedge (b \vee b') = \top \wedge (b \vee b') = b \vee b'.$$

Similarly $b' = b \vee b'$ so that $b = b'$.

In a Heyting algebra H the pseudocomplement a^* of an element a is not, in general, a complement for a. (Consider the Heyting algebra of open sets of a topological space.) But there is a simple necessary and sufficient condition on a Heyting algebra for all pseudocomplements to be complements: this is stated in the following

Proposition. *The following conditions on a Heyting algebra H are equivalent:*

(i) *pseudocomplements are complements, i.e. $x \vee x^* = \top$ for all $x \in H$;*
(ii) *pseudocomplementation is of order 2, i.e. $x^{**} = x^*$ for all $x \in H$.*

Proof. (i) → (ii). Assuming (i), we have

$$x^{**} = x^{**} \wedge \top = x^{**} \wedge (x \vee x^*) = (x^{**} \wedge x) \vee (x^{**} \wedge x^*) = (x^{**} \wedge x) \vee \bot = (x^{**} \wedge x).$$

Therefore $x^{**} \leq x$ whence $x^{**} = x$.

(ii)→ (i). We have $(x \vee x^*)^* = x^* \wedge x^{**} = \bot$, so assuming (ii) gives $x \vee x^* = (x \vee x^*)^{**} = \bot^* = \top$. ∎

We now define a *Boolean algebra* to be a Heyting algebra satisfying either of the equivalent conditions (i) or (ii). The following identities accordingly hold in any Boolean algebra:

$$x \vee y = y \vee x, \quad x \wedge y = y \wedge x$$
$$x \vee (y \vee z) = (x \vee y) \vee z, \quad x \wedge (y \wedge z) = (x \wedge y) \wedge z$$
$$(x \vee y) \wedge y = y, \quad (x \wedge y) \vee y = y$$
$$x \wedge (y \vee z) = (x \wedge y) \vee (x \wedge z), \quad x \vee (y \wedge z) = (x \vee y) \wedge (x \vee z).$$
$$x \vee x^* = \top, \quad x \wedge x^* = \bot.$$
$$(x \vee y)^* = x^* \wedge y^*, \quad (x \wedge y)^* = x^* \vee y^*$$
$$x^{**} = x$$

It is easy to show that in any Boolean algebra $x \Rightarrow y = x^* \vee y$. In a *complete* Boolean algebra we have the following identities:

$$(\bigvee_{i \in I} x_i)^* = \bigwedge_{i \in I} x_i^* \quad (\bigwedge_{i \in I} x_i)^* = \bigvee_{i \in I} x_i^* \quad x \wedge (\bigvee_{i \in I} y_i)^* = \bigvee_{i \in I}(x \wedge y_i)$$
$$x \wedge \bigwedge_{i \in I} y_i = \bigwedge_{i \in I}(x \vee y_i).$$

Calling a lattice *complemented* if it is bounded and each of its elements has a complement, we can characterize Boolean algebras alternatively as *complemented distributive lattices*. For we have already shown that every Boolean algebra is distributive and complemented. Conversely, given a complemented distributive lattice L, write a^c for the (unique) complement of an element a; it is then easily shown that defining implication by $x \Rightarrow y = x^c \vee y$ turns L into a Heyting algebra in which x^* coincides with x^c, so that L is Boolean.

The meet, join, and complementation operations in a Boolean algebra are called its *Boolean operations*. A *subalgebra* of a Boolean algebra B is a nonempty subset closed under B's Boolean operations. Clearly a subalgebra of a Boolean algebra B is itself a Boolean algebra with the same top and bottom elements as those of B.

Examples of Boolean algebras.

(i) The linearly ordered set $2 = \{0, 1\}$ with $0 < 1$ is a complete Boolean algebra, the *2-element algebra*.

(ii) The power set lattice PA of any set A is a complete Boolean algebra. A subalgebra of a power set algebra is called a *field of sets*.

(iii) Let $F(A)$ consist of all finite subsets and all complements of finite subsets of a set A. With the partial ordering of inclusion, $F(A)$ is a field of sets called the *finite-cofinite algebra* of A.

(iv) Let X be a topological space, and let $C(X)$ be the family of all simultaneously closed and open ("clopen") subsets of X. With the partial ordering of inclusion, $C(X)$ is a Boolean algebra called the *clopen algebra* of X.

An element a of a Heyting algebra H is said to be *regular* if $a = a^{**}$. Clearly a Heyting algebra is a Boolean algebra if and only if each of its elements is regular. Let B be the set of regular elements of H; it can be shown that B, with the partial ordering inherited from H, is a Boolean algebra in which the operations $\wedge$ and $*$

coincide with those of H, but[33] $\vee_B = (\vee_H)^{**}$. If H is complete, so is B; the operation $\wedge$ in B coincides with that in H while $\vee_B = (\vee_H)^{**}$. B is written H^{bool} and called the *Booleanization* of H.

COVERAGES AND THEIR ASSOCIATED FRAMES

Let $(P, \leq)$ be a fixed but arbitrary partially ordered set: we shall use letters p, q, r, s, t to denote elements of p. A subset S of P is said to be a *sharpening of*, or to *sharpen*, a subset T of P, written $S \prec T$, if $\forall s \in S \exists t \in T(s \leq t)$. A *sieve* in P is a subset S such that $p \in S$ and $q \leq p$ implies $q \in S$. Each subset S of P generates a sieve $\overline{S}$ given by $\overline{S} = \{p : \exists s \in S(p \leq s)\}$.

A *coverage* on P is a map $\mathbf{C}$ assigning to each $p \in P$ a family $\mathbf{C}(p)$ of subsets of $p\!\downarrow\, = \{q: q \leq p\}$, called *(C-)covers of p*, such that, if $q \leq p$, any cover of p can be sharpened to a cover of q, i.e.,

(*) $$S \in \mathbf{C}(p) \,\&\, q \leq p \to \exists T \in \mathbf{C}(q)[\forall t \in T \exists s \in S(t \leq s)].$$

Now we associate a *frame* with each coverage $\mathbf{C}$ on P. First, we define $\hat{P}$ to be the set of sieves in P partially ordered by inclusion: $\hat{P}$ is then a frame — the *completion* of P — in which joins and meets are just set-theoretic unions and intersections, and in which the operations $\Rightarrow$ and $\neg$ are given by

$$I \Rightarrow J = \{p : I \cap p\!\downarrow\, \subseteq J\} \qquad \neg I = \{p : I \cap p\!\downarrow\, = \emptyset\}.$$

Given a coverage $\mathbf{C}$ on P, a sieve I in P is said to be *C-closed* if

$$\exists S \in \mathbf{C}(p)(S \subseteq I) \to p \in I.$$

We write H_C for the set of all **C**-closed sieves in P, partially ordered by inclusion.

---

[33] Here we write $\vee_B, \vee_H$ for the join operations in B and H.

Lemma. *If $I \in H_C$, $J \in H_C$, then $I \Rightarrow J \in H_C$.*

Proof. Suppose that $I \in \hat{P}$, $J \in H_C$ and $S \subseteq I \Rightarrow J$ with $S \in C(p)$. Define $U = \{q \in I : \exists s \in S.\ q \leq s\}$. Then $U \subseteq J$. If $q \in I \cap p\!\downarrow$, then there is $T \in C(q)$ for which $T \prec S$. Then for any $t \in T$, there is $s \in S$ for which $t \leq s$, whence $t \in U$. Accordingly $T \subseteq U \subseteq J$. Since J is C-closed, it follows that $q \in J$. We conclude that $I \cap p\!\downarrow \subseteq J$, whence $p \in p\!\downarrow \subseteq I \Rightarrow J$. Therefore $I \Rightarrow J$ is C-closed. ∎

It follows from the lemma that H_C is a frame. For clearly an arbitrary intersection of C-closed sieves is C-closed. So H_C is a complete lattice. In view of the lemma the implication operation in $\hat{P}$ restricts to one in H_C making H_C a Heyting algebra, and so a frame. H_C is called the frame *associated* with C.

CONNECTIONS WITH LOGIC.

Heyting and Boolean algebras have close connections with *intuitionistic* and *classical* logic[34], respectively.

Intuitionistic first-order logic has the following axioms and rules of inference.

Axioms

$$\alpha \to (\beta \to \alpha)$$
$$[\alpha \to (\beta \to \gamma)] \to [(\alpha \to \beta) \to (\alpha \to \gamma)]$$
$$\alpha \to (\beta \to \alpha \wedge \beta)$$

$\alpha \wedge \beta \to \alpha \qquad\qquad \alpha \wedge \beta \to \beta$
$\alpha \to \alpha \vee \beta \qquad\qquad \beta \to \alpha \vee \beta$

$$[\alpha \to (\beta \to \gamma)] \to [(\alpha \to \beta) \to (\alpha \to \gamma)]$$
$$(\alpha \to \gamma) \to [(\beta \to \gamma) \to (\alpha \vee \beta \to \gamma)]$$
$$(\alpha \to \beta) \to [(\alpha \to \neg\beta) \to \neg\alpha]$$
$$\neg\alpha \to (\alpha \to \beta)$$
$$\alpha(t) \to \exists x \alpha(x) \qquad \forall x \alpha(x) \to \alpha(y) \quad (x \text{ free in } \alpha \text{ and } t \text{ free for } x \text{ in } \alpha)$$

$$x = x \qquad \alpha(x) \wedge x = y \to \alpha(y)$$

---

[34] For accounts of both systems of logic, see, e.g. Bell and Machover [1977].

Rules of Inference

$$\frac{\alpha,\ \alpha \to \beta}{\beta}$$

$$\frac{\beta \to \alpha(x)}{\beta \to \forall x \alpha(x)} \qquad \frac{\alpha(x) \to \beta}{\exists x \alpha(x) \to \beta}$$

(x not free in β)

Classical first-order logic is obtained by adding to the intuitionistic system the rule of inference

$$\frac{\neg\neg\alpha}{\alpha}$$

In intuitionistic logic none of the classically valid logical schemes

- **LEM** (law of excluded middle) $\quad \alpha \vee \neg\alpha$
- **LDN** (law of double negation) $\quad \neg\neg\alpha \to \alpha$
- **DML** (de Morgan's law) $\quad \neg(\alpha \wedge \beta) \to \neg\alpha \vee \neg\beta$

are derivable. However **LEM** and **LDN** are intuitionistically equivalent and **DML** is intuitionistically equivalent to the weakened law of excluded middle:

WLEM $\quad \neg\alpha \vee \neg\neg\alpha$.

Also the weakened form of **LDN** for negated statements,

WLDN $\quad \neg\neg\neg\alpha \to \neg\alpha$

is intuitionistically derivable. It follows that any formula intuitionistically equivalent to a negated formula satisfies **LDN**.

Heyting algebras are associated with theories in intuitionistic logic in the following way. Given a consistent theory T in an intuitionistic propositional or first-order language $\mathscr{L}$, define the equivalence relation $\approx$ on the set of formulas of $\mathscr{L}$ by $\varphi \approx \psi$ if $T \vdash \varphi \leftrightarrow \psi$. For each formula φ write $[\varphi]$ for its $\approx$-equivalence class. Now define the relation $\leq$ on the set $H(T)$ of $\approx$-equivalence classes by $[\varphi] \leq [\psi]$ if and only if $T \vdash \varphi \to \psi$. Then $\leq$ is a partial ordering of $H(T)$ and the partially ordered set $(H(T), \leq)$ is a Heyting algebra in which $[\varphi] \Rightarrow [\psi] = [\varphi \to \psi]$, with analogous equalities defining the meet and join operations, 0, and 1. $H(T)$ is called the Heyting algebra *determined by* T. It can be shown that Heyting algebras of the form $H(T)$ are typical in the sense that, for any Heyting algebra L, there is a propositional intuitionistic theory T such that L is isomorphic to $H(T)$. Accordingly Heyting algebras may be identified as the algebras of intuitionistic logic.

Similarly, starting with a consistent theory T in a classical propositional or first-order language, the associated algebra $B(T)$ is a Boolean algebra known as the *Lindenbaum algebra* of T. Again, it can be shown that any Boolean algebra is isomorphic to $B(T)$ for a suitable classical theory T.

As regards semantics, Heyting algebras and Boolean algebras have corresponding relationships with intuitionistic, and classical, propositional logic, respectively. Thus, suppose given a propositional language; let $\mathscr{P}$ be its set of propositional variables. Given a map $f: \mathscr{P} \to H$ to a Heyting algebra H, we extend f to a map $\alpha \mapsto [\![\alpha]\!]$ of the set of formulas of $\mathscr{L}$ to H by:

$$[\![\alpha \land \beta]\!] = [\![\alpha]\!] \land [\![\beta]\!] \quad [\![\alpha \lor \beta]\!] = [\![\alpha]\!] \lor [\![\beta]\!] \quad [\![\alpha \Rightarrow \beta]\!] = [\![\alpha]\!] \Rightarrow [\![\beta]\!] \quad [\![\neg \alpha]\!] = [\![\alpha]\!]^*$$

A formula α is said to be *Heyting valid* — written $\models \alpha$ — if $[\![\alpha]\!] = \top$ for any such map f. It can then be shown that α is Heyting valid iff $\vdash \alpha$ in the intuitionistic propositional calculus, i.e., iff α is provable from the propositional axioms listed above.

Similarly, if we define the notion of Boolean validity by restricting the definition of Heyting validity to maps into Boolean algebras, then it can be shown that a formula is Boolean valid iff it is provable in the classical propositional calculus.

Finally, again as regards semantics, *complete* Heyting and Boolean algebras are related to intuitionistic, and classical *first-order* logic, respectively. To be precise, let $\mathscr{L}$ be a first-order language whose sole extralogical symbol is a binary predicate symbol P. A *Heyting-valued ℒ-structure* is a quadruple **M** = (M, eq, Q, H), where M is a nonempty set, H is a complete Heyting algebra and eq and Q are maps $M^2 \to M$ satisfying, for all $m, n, m', n' \in M$,

$$eq(m, m) = \top, \quad eq(m, n) = eq(n, m), \quad eq(m, n) \wedge eq(n, n') \leq eq(m, n'),$$
$$Q(m, n) \wedge eq(m, m') \leq Q(m', n), \quad Q(m, n) \wedge eq(n, n') \leq Q(m, n').$$

For any formula α of ℒ and any finite sequence $x = \langle x_1, ..., x_n \rangle$ of variables of ℒ containing all the free variables of α, we define for any Heyting-valued ℒ-structure **M** a map

$$[\![\alpha]\!]^{\mathbf{M}}_x : M^n \to H$$

recursively as follows:

$$[\![x_p = x_q]\!]^{\mathbf{M}}_x = \langle m_1, ..., m_n \rangle \mapsto eq(m_p, m_q),$$
$$[\![Px_p x_q]\!]^{\mathbf{M}}_x = \langle m_1, ..., m_n \rangle \mapsto Q(m_p, m_q),$$
$$[\![\alpha \wedge \beta]\!]^{\mathbf{M}}_x = [\![\alpha]\!]^{\mathbf{M}}_x \wedge [\![\beta]\!]^{\mathbf{M}}_x,$$

and similar clauses for the other connectives,

$$[\![\exists y\, \alpha]\!]^{\mathbf{M}}_x = \langle m_1, ..., m_n \rangle \mapsto \bigvee_{m \in M} [\![\alpha\,(y/u)]\!]^{\mathbf{M}}_{ux}(m, m_1, ..., m_n)$$
$$[\![\forall y\, \alpha]\!]^{\mathbf{M}}_x = \langle m_1, ..., m_n \rangle \mapsto \bigwedge_{m \in M} [\![\alpha\,(y/u)]\!]^{\mathbf{M}}_{ux}(m, m_1, ..., m_n)$$

Call α **M**-*valid* if $[\![\alpha]\!]^{\mathbf{M}}_x$ is identically ⊤, where x is the sequence of all free variables of α. Then it can be shown that α is **M**-*valid for all* **M** *iff* α *is provable in intuitionistic first-order logic*. This is the *algebraic completeness theorem* for intuitionistic first-order logic.

Similarly, if we carry out the same procedure, replacing complete Heyting algebras with complete Boolean algebras, one can prove the corresponding algebraic completeness theorem for classical first-order logic, namely, a first-order formula is valid in every Boolean-valued structure iff it is provable in classical first-order logic.

Concluding Observations

In this book we have used frame-valued universes in proving the consistency of set-theoretic assertions with **IST**. In fact, what has become the standard procedure for proving the consistency with **IST** of a given assertion p is to construct a certain sort of category - a *topos*[35] - in which p holds in a "natural" sense.

For example, the subcountability of $\mathbb{N}^\mathbb{N}$, as well as many of the other assertions concerning $\mathbb{N}$ mentioned in the Introduction, can be shown to hold in the so-called *effective topos* **Eff** In **Eff** maps between objects constructed from the natural numbers correspond to (partial) *recursive functions* between them. In particular the countable subsets of $\mathbb{N}$ may be identified with the recursively enumerable subsets, and the detachable subsets of $\mathbb{N}$ with the recursive subsets. The object $\mathbb{N}^\mathbb{N}$ may be considered as the set **Rec** of (total) recursive functions on $\mathbb{N}$ [36]. That being the case, if we write φ_x for the partial recursive function on $\mathbb{N}$ with index x, and U for the set of indices of total recursive functions[37], the map $x \mapsto \varphi_x$ for $x \in U$ is, in **Eff**, a surjection from U to $\mathbb{N}^\mathbb{N}$, making $\mathbb{N}^\mathbb{N}$ subcountable in **Eff**. On the other hand $\mathbb{N}^\mathbb{N}$ is not numerable in **Eff** because it can be shown that, in **Eff**, *Brouwer's Continuity Principle* holds. This is the assertion

$$\forall F \in \mathbb{N}^{(\mathbb{N}^\mathbb{N})} \forall f \in \mathbb{N}^\mathbb{N} \exists n \in \mathbb{N} \forall g \in \mathbb{N}^\mathbb{N} [\forall m < n(f(m) = g(m)) \to F(f) = F(g)]$$ [38]

---

[35] Accounts of topos theory may be found in Bell [1988], Goldblatt [1979], Johnstone [1977] and [2002], Lambek and Scott [1986], Mac Lane and Moerdijk [1992] and McLarty [1992].

[36] Hence, in **Eff**, *Church's thesis* holds in the strong sense that *every* function $\mathbb{N} \to \mathbb{N}$ is recursive.

[37] Since $\mathbb{N}^\mathbb{N}$ is uncountable, so must U be; this corresponds to the result from recursion theory that the set of indices of total recursive functions is not recursively enumerable.

[38] In **Eff** the set $\mathbb{N}^{(\mathbb{N}^\mathbb{N})}$ of all maps $\mathbb{N}^\mathbb{N} \to \mathbb{N}$ may be identified with the set of *effective operations*, that is, the set of functions $F: \mathbf{Rec} \to \mathbb{N}$ such that there is a partial recursive function φ with the property that for every $f \in \mathbf{Rec}$ and every index n for f, we have $\varphi(n) = F(f)$. Brouwer's continuity principle asserts the continuity of every map to $\mathbb{N}$ from Baire space $\mathcal{N}$.

From this it would follow that, if F were an injection from $\mathbb{N}^\mathbb{N}$ to $\mathbb{N}$, then, for each $f \in \mathbb{N}^\mathbb{N}$ there would exist an $n \in \mathbb{N}$ for which

$$\forall g \in \mathbb{N}^\mathbb{N} \, \forall m < n(f(m) = g(m)) \to f = g),,$$

which is clearly impossible. Thus $\mathbb{N}^\mathbb{N}$ is not numerable in **Eff**, and hence nofr is Par*($\mathbb{N}, \mathbb{N}$), since it contains $\mathbb{N}^\mathbb{N}$.

In **Eff**, both Par*($\mathbb{N}, \mathbb{N}$) and **P***$\mathbb{N}$ are countable. That Par*($\mathbb{N}, \mathbb{N}$) is countable in Eff follows from the fact that it corresponds to the set of *partial recursive functions* on $\mathbb{N}$. The map $x \mapsto \varphi_x$ assigning to each $x \in \mathbb{N}$ the partial recursive function with index x is, in **Eff**, a surjection from $\mathbb{N}$ to Par*($\mathbb{N}, \mathbb{N}$). (That being the case, as we noted above, the set $\{x \in \mathbb{N} : x \notin \text{dom}(\varphi_x)\}$ must be (in **Eff**) *uncountable*. This corresponds to the fact that this set is not recursively enumerable.) A similar argument – using the fact that **P***$\mathbb{N}$ in **Eff** corresponds to the set of recursively enumerable subsets of $\mathbb{N}$ - shows that **P***$\mathbb{N}$ is also countable in **Eff**.

While $\mathbb{N}^\mathbb{N}$ fails to be numerable in **Eff**, Bauer [2011] has shown it to be numerable in the related topos **Eff!** in which maps between objects constructed from the natural numbers correspond to functions which are *infinite time computable*, that is, computable by an infinite time Turing machine. This is a Turing machine which is allowed to run *infinitely long*, with the computation steps counted by ordinals. The power of these machines far exceeds that of ordinary Turing machines: for example, both the halting problem and the problem of deciding the equality of two total recursive functions are soluble using infinite time machines. In **Eff!**, just as in **Eff**, $\mathbb{N}^\mathbb{N}$ is subcountable. But in **Eff!** $\mathbb{N}^\mathbb{N}$ also satisfies the *axiom of choice* in the form: any total relation defined on $\mathbb{N}^\mathbb{N}$ contains a function (this cannot be the casein **Eff**). Putting these two facts together quickly yields an injection of $\mathbb{N}^\mathbb{N}$ into $\mathbb{N}$. We conclude that the numerability of $\mathbb{N}^\mathbb{N}$ is consistent with **IST**.

There are a number of topos models of Brouwer's Principle (**BP**) that all real functions are continuous. As we have essentially shown in Chapter IV, **BP** holds in **Shv**($\mathcal{A}$)[39]. Mac Lame and Moerdijk [1992] present a different kind of

---

[39] In fact **BP** can be shown to hold in many other spatial toposes: see Hyland [1979]

topos model of **BP**. **BP** also holds in any of the so-called *smooth* toposes: see Bell [2008], McLarty [1992] and Moerdijk and Reyes [1991]. A smooth topos may be considered to be an enlargement of the category **Man** of manifolds (or spaces) and smooth maps to a topos which contains no new maps between spaces, so that all such maps there – in particular those from $\mathbb{R}$ to $\mathbb{R}$ are still smooth, and so *a fortiori* continuous.

Historical Notes

Chapter I. Friedman [1973i, 1973ii] and Myhill [1973] seem to have been the first to investigate systems of intuitionistic set theory. Crayson [1978] undertakes a systematic investigation of topology and ordinal arithmetic in an intuitionistic setting. That **LEM** follows from the axiom of choice was first proved, in a category-theoretic setting, by Diaconescu [1975]; the logical version was formulated and proved by Goodman and Myhill [1978]. The investigation of the connection between choice principles and logical principles is taken from Bell [2006}; see also Bell [2009].

Chapter II. The characterization of $\mathbb{N}$ in terms of the simple recursion principle (Proposition 5) is due to F. W. Lawvere in a category-theoretic setting. Propostion 11 concerning monics on Ω is due to Denis Higgs. Work on finite sets in an intuitionistic setting (or their equivalents, finite objects in a topos), has been extensive: for a complete bibliography see Johnstone [2002]. The section on Frege's theorem is taken from Bell [1999i] and [1991ii].

Chapter III. Much of the discussion of real numbers presented here is based on Johnstone's [2002] account of real numbers in a topos. Proposition 2 is due to Johnstone [1979].

Chapter IV. Frame-valued models were first investigated by Grayson [1975] and [1979], where it is also shown that Zorn's lemma is consistent with **IZF** (see also Bell [1997]). The consistency of **ZF** relative to **IZF** was first proved by Friedman [1973] and Powell [1975]; the proof given here is due to Grayson [1979]. A topos model in which $\mathbb{N}^\mathbb{N}$ is subcountable was first produced by A. Joyal (see Fourman and Hyland [1979], Johnstone [2002]). The model of the subcountability of $\mathbb{N}^\mathbb{N}$ given in the text is a frame-valued version of the topos presented in Example D.4.1.9. of Johnstone [2002]. The representation of real numbers (in a sheaf topos) is due to M. Tierney. The proof that Brouwer's Principle holds for the real numbers over Baire space is due to Scott [1970]. The status of Brouwer's Principle in spatial toposes has been investigated by Hyland [1979]. The failure of **FTA** in the sheaf topos over the complex numbers was first noted in Fourman and Hyland [1979].

Concluding Observations. The effective topos was first introduced by Hyland [1983]. The concept of a smooth topos is due to F. W. Lawvere.

Bibliography

Bauer, A. [2011]. An injection from $\mathbb{N}^{\mathbb{N}}$ to $\mathbb{N}$. http://math.andrej.com/wp-content/uploads/2011/06/injection.pdf

Bell, J. L. [1988]. *Toposes and Local Set Theories: An Introduction.* Clarendon Press, Oxford, 1988. Dover reprint 2007.

-------- [199i] Frege's theorem in a constructive setting. *J. Symbolic Logic* 64, no. 2, 486-488.

-------- [199ii]. Finite sets and Frege structures *J. Symbolic Logic*, 64, no. 4, 152-156.

---------[1997]. Zorn's lemma and complete Boolean algebras in intuitionistic type theories. *J. Symbolic Logic.* **62**, 1265-1279.

---------[2006]. Choice principles in intuitionistic set theory. In *A Logical Approach to Philosophy.* Springer. Heidelberg-London, New York.

---------[2008]. *A Primer of Infinitesimal Analysis,* 2nd. edition. Cambridge University Press.

-------- [2009]. *The Axiom of Choice.* College Publications, London.

---------[2011]. *Set Theory: Boolean-valued Models and Independence Proofs.* 3rd edition. Clarendon Press, Oxford.

Bell, J. L. and Machover, M. [1977]. *A Course in Mathematical Logic.* North-Holland, Amsterdam

Diaconescu, R. [1975]. Axiom of choice and complementation. *Proc. Amer. Math. Soc.* **51**, 176-8.

Fourman, M.P. and Hyland, J.M.E. [1979]. Sheaf models for analysis. In Fourman, M. P., Mulvey, C. J., and Scott, D. S. (eds.) *Applications of Sheaves. Proc.*

L.M.S. Durham Symposium 1977. Springer Lecture Notes in Mathematics 753, pp. 280 – 301.

Fourman, M.P. and Scptt, D.S. [1979]. Sheaves and logic. In Fourman, M. P., Mulvey, C. J., and Scott, D. S. (eds.) *Applications of Sheaves. Proc. L.M.S. Durham Symposium* 1977. Springer Lecture Notes in Mathematics 753, pp. 302-401.

Friedman, H. [1973i]. Some applications of Kleene's methods for intuitionistic systems. in A.R.D. Mathia, A. and Rogers, H. (eds.) *Proceedings of the 1971 Cambridge Summer School in Mathematical Logic* Springer Lecture Notes in Mathematics 337, pp. 113–170.

---------[1973ii]. The consistency of classical set theory relative to a set theory with intuitionistic logic. *Journal of Symbolic Logic*, 38: 315–319.

Goldblatt, R. [1979]. *Topoi: The Categorial Analysis of Logic*. North-Holland, Amsterdam.

Goodman, N. and Myhill, J. [1978]. Choice implies excluded middle. *Z. Math Logik Grundlag. Math* **24**, no. 5, 461.

Grayson, R.J. [1975]. *A sheaf approach to models of set theory*. M.Sc. thesis, Oxford University.

---------[1978]. *Intuitionistic Set Theory*. D.Phil. Thesis, Oxford University.

---------[1979.]. Heyting-valued models for intuitionistic set theory. In Fourman, M. P., Mulvey, C. J., and Scott, D. S. (eds.) *Applications of Sheaves. Proc. L.M.S. Durham Symposium* 1977. Springer Lecture Notes in Mathematics 753, pp. 402-414.

Halmos, P. R. [1963]. *Lectures on Boolean Algebras*. Van Nostrand, New York.

Higgs, D. [1973]. A category approach to Boolean-valued set theory. Lecture Notes, University of Waterloo.

Hyland, J. M. E. [1979]. Continuity in spatial toposes. In Fourman, M. P., Mulvey, C. J., and Scott, D. S. (eds.) *Applications of Sheaves. Proc. L.M.S. Durham Symposium* 1977. Springer Lecture Notes in Mathematics 753, pp. 442-465.

---------[1983]. The effective topos. In *The L. E. J. Brouwer Centenary Symposium*, Studies in Logic and the Foundations of Math., vol. 110, North Holland, Amsterdam, pp. 165-216.

Johnstone, P. T. [1977]. *Topos Theory*. Academic Press, London.

---------[1979]. Conditions related to De Morgan's law. In Fourman, M. P., Mulvey, C. J., and Scott, D. S. (eds.) *Applications of Sheaves. Proc. L.M.S. Durham Symposium* 1977. Springer Lecture Notes in Mathematics 753, pp. 479-491.

---------[2002]. *Sketches of an Elephant: A Topos Theory Compendium*, vols. I and II. Oxford Logic Guides vols. 43 and 44. Clarendon Press, Oxford.

Lambek, J. and Scott, P. J. [1986]. *Introduction to Higher-Order Categorical Logic*. Cambridge University Press.

Mac Lane, S. and Birkhoff, G. [1967]. *Algebra*. Macmillan, New York.

Mac Lane, S. and Moerdijk, I. [1992]. *Sheaves in Geometry and Logic: A First Introduction to Topos Theory*. Springer-Verlag, Berlin.

McLarty. C. [1992]. *Elementary Categories, Elementary Toposes*. Clarendon Press, Oxford.

Moerdijk, I. and Reyes, G.E. [1991]. *Models for Smooth Infinitesimal Analysis*. Springer-Verlag, Berlin.

Myhill, J. [1973]. Some properties of Intuitionistic Zermelo-Fraenkel set theory. In Mathias, A. and Rogers, H. (eds.) *Proceedings of the 1971 Cambridge Summer School in Mathematical Logic* Springer Lecture Notes in Mathematics 337, pp. 206-231.

Powell, W. [1975]. Extending Gödel's negative interpretation to ZF. *Journal of Symbolic Logic*, 40: 221–229.

Scott, D. S. [1970]. Extending the topological interpretation to intuitionistic analysis, II. . In Myhill, J. Kino, A. and Vesley, R. E. , eds. , *Intuitionism and Proof Theory*, North-Holland, Amsterdam, pp. 235-255.

Index

associated frame 108
Axiom of Choice 20
Axiom of Choice for I 83
Axioms of **IZ** 12, 13
Axioms of **IZF** 55

Baire space 95
Booleanization 107
Brouwer's Principlde 95

Cantor's Theorem 2, 5
Cauchy real number 53
Cauchy sequence 53
cohesive 8, 97
Collection argument 62
complemented 65
completely $\bot$, ($\bot$, 2) distributive 77
conditionally order-complete 50
connected 74
continuum 7
core 67
correlated 88, 89, 92
countable 1
Countable Axiom of Choice 83
countably generated 84
cover 107
coverage 107
cumulative hierarchy 58

Dedekind finite 38
Dedekind infinite 36
Dedekind real number 48
definite element 62
Definite Element Lemma 65
degree 85

detachable 17
discrete 17
disjoint refinement 72
distributive, $\bot$-(I, J) 75
DML 16

Existence Principle 72
Extended Recursion Principle 34
extensional 58

false 60
finite 38
frame 104
frame of truth values 16
frame-valued model 60
frame-valued universe 61
Frege structure/strict 41
Frege's Theorem 40

H-extension 60
H-valued model of **IZF** 67
H-valued real number 85
H-valued structure 59

indecomposable 8
Induction Principle for Peano structure 29
Induction Principle for $V^{(H)}$ 60
inductive 27
inhabited 13, 62
intuitioistic set theory **IST** 1

Kuratowski finite 38

LEM 1, 16

local 91
localizable 92
locally connected 75

Mixing Lemma 64

natural numbers 27
near-local 91
numerable 1

ordered pair 13, 67

Peano structure 29
Peano's axioms 29
preserves exponentials 77
Principle of Induction on Ordinals 56
pseudocomplement 102

rank function 58
real function over X 91
real number over X, U 86, 87
refinable 72
Refinable Existence Lemma 73
refinement 72
regular element 106

set of generators 84
sieve 107
Simple Recursion Principle 30
spatial extension 60
strictly finite 38
strong core 72
subcountable 1
subquotient 74
subset classifier 16

totally disconnected 65
transitive 58
true 6
truth value 60

Unique Existence Principle for $\aleph^{(H)}$ 64
Universe of H-sets 61

weak real number 49
well-founded 57
well-ordering 58
WLEM 16

Zorn's Lemma 69

www.ingramcontent.com/pod-product-compliance
Ingram Content Group UK Ltd.
Pitfield, Milton Keynes, MK11 3LW, UK
UKHW021321180426
11947UKWH00015B/1358